# Evolution, Development, &
# The Predictable Genome

# Evolution, Development, & The Predictable Genome

## David L. Stern

ROBERTS AND COMPANY PUBLISHERS

Greenwood Village, Colorado

**Roberts and Company Publishers**
4950 South Yosemite Street, F2 #197
Greenwood Village, Colorado 80111 USA
Internet: www.roberts-publishers.com
Telephone: (303) 221-3325
Facsimile: (303) 221-3326
E-mail: info@roberts-publishers.com

Publisher: Ben Roberts
Copyeditor: Gunder Hefta
Production Manager: Mark Ong at Side By Side Studios
Cover and Interior Designer: Mark Ong at Side By Side Studios
Cover art: Laura Zindel

Library of Congress Cataloging-in-Publication Data
100 3656688

Stern, David Lawrence, 1966-
 Evolution, development, & the predictable genome / David L. Stern.
    p. cm.
 Includes bibliographical references and index.
 ISBN 978-1-936221-01-1
 1. Evolutionary genetics. 2. Developmental biology. I. Title.
 QH390.S74 2011
 572.8'38--dc22
                                        2009050542

10 9 8 7 6 5 4 3 2 1

*This book is dedicated to*

*Paul Sherman*
*Charles Aquadro*
*Carlos Martinez del Rio*
*&*
*Michael Akam*

*each of whom, in his own way, transformed my understanding of*
*biological diversity*

The only objections that have occurred to me are
1st that you have loaded yourself with an
unnecessary difficulty in adopting 'Natura non facit
saltum' so unreservedly. I believe she does make
*small* jumps—and 2nd it is not clear to me why if
external physical conditions are of so little moment
as you suppose variation should occur at all.

—Thomas H. Huxley, in letter to Charles Darwin, upon
finishing *The Origin of Species*, November 23, 1859

You have most cleverly hit on one point, which has
greatly troubled me; if, as I must think external
conditions produce little *direct* effect, what the devil
determines each particular variation. What makes a
tuft of feathers come on a Cock's head; or moss on
a moss-rose?—I shall much like to talk over this
with you.

—Darwin's reply to Huxley, November 25, 1859

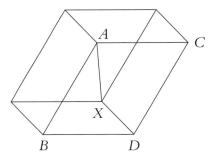

The rhomboid $AX$ is drawn so that the solid angle $A$ should be seen the nearest to the spectator, and the solid angle $X$ the furthest from him, and that the face $ACBD$ should be the foremost, while the face $XDC$ is behind. But in looking repeatedly at the same figure, you will perceive that at times the apparent position of the rhomboid is so changed that the solid angle $X$ will appear the nearest, and the solid angle $A$ the furthest; and that the face $ACDB$ will recede behind the face $XDC$, which will come forward; which effect gives to the whole solid a quite contrary apparent inclination.... After... attentive analysis of the fact, it occurred to me, that it was owing to an involuntary change in the adjustment of the eye for obtaining distinct vision.

—Prof. L. A. Necker, "Observations on some remarkable Optical Phaenomena seen in Switzerland; and on an optical Phaenomenon which occurs on viewing a Figure of a Crystal or geometrical Solid."

# CONTENTS

# PREFACE

When a population evolves in response to natural selection, organisms change—in appearance, physiology, or behavior—over time. Colder temperatures select for thicker fur. A new pathogen selects for pathogen resistance. Faster predators select for swifter prey. Since individuals vary for many characteristics—fur thickness, pathogen resistance, running speed—some will survive and reproduce better than others in the face of ecological challenges. These "fitter" individuals contribute disproportionately to the next generation. If the differences between individuals have a genetic basis, then more individuals of the next generation will share the characteristics—thick fur, robust immune systems, fleet feet—that helped their parents to survive and to reproduce. The population has evolved.

Differences between individuals—the raw material of evolution—are caused by variation in genes and by variation in the environment. The genetic component of the variation starts as mutations. Since each aspect of an organism—fur, microbial defenses, limbs—is constructed by multiple genes, and since mutations occur randomly in the genome, we might expect that a diverse set of mutations, occurring in many of these genes, would be selected for during evolution, but this is not the case. Natural selection favors some kinds of mutations and disfavors others. This is because mutations alter organisms by changing the way that genes act during development. Some mutations may cause a single specific change in development, whereas others may cause multifarious changes. Natural selection tends to favor mutations that result in limited, specific changes. Thus, development will bias which mutations are selected.

But development is only half the story. All evolution occurs within populations, and different kinds of mutations will be selected in different kinds of populations. For example, a mutation causing a small change will be more efficiently selected in a large population than in a small population. Other aspects of population biology also influence how selection acts. Evolutionary mechanisms, like developmental mechanisms, bias which mutations are selected. Therefore, in seeking to understand how genomes evolve to generate biological diversity, we should consider both development and mechanisms of evolution.

This book presents one way to integrate evolutionary biology with developmental biology. Many other attempts at integrating evolution and development have been made, particularly in the past few decades. In contrast to most of these previous efforts, I do not discuss broad-scale patterns across large spans of evolutionary history. Instead, I focus on the individual steps of evolutionary change, as revealed by examination of variation within species and between closely related species. As William Bateson wrote in 1900, in an attempt to promote the new discipline he later called genetics: "The essential problem of evolution is how any one given step in evolution was accomplished." I explore how the mechanisms that generate each step in evolution—the process that ultimately generates differences between species—interact with development. I focus on the genetic causes of evolution and therefore rely heavily on insights from population genetics. Currently, population genetics and developmental biology are about as far from integrated as any two biological disciplines could be. This intellectual chasm probably has many causes, and it is worth comparing several salient differences between the disciplines.

Population genetics provides a framework for attempting to understand how, looking forward, the mechanisms of evolution will influence the genetic fate of populations and how, looking back, patterns observed in modern populations are best explained by the action of evolutionary mechanisms in the past. Population genetics is a bit like weather prediction. Sometimes the starting conditions can be defined precisely enough to allow accurate short-term prediction, but long-term prediction is complicated by the intercession of random events. Nonetheless, population genetics provides a guide to likely evolutionary outcomes. Similarly, meteorologists cannot say for certain where every hurricane will make landfall, but they can predict when during the year hurricanes will be prevalent and the general geographic areas the hurricanes will hit. But population geneticists rarely spend much time trying to predict the future. Population genetics is mainly about the past: What historical forces have generated current patterns of genetic variation? Are current patterns best explained by the deterministic force of natural selection, or by random processes?

Development, by contrast, is only weakly influenced by random events. Of course, random events cause some variation in development, but the scale of this variation pales in comparison to the variation caused by unpredictable events in population genetics.

Population genetics and developmental biology employ dramatically different explanatory frameworks. Explanation in population genetics proceeds from indirect inference of the causes of evolution through statistical analysis of data collected from populations. Explanation in developmental biology usually involves experiments that provide reasonably direct evidence for the causes of development.

At first glance, then, attempts to integrate development and population genetics would seem to make little sense. What benefit could possibly come from trying to integrate a very noisy process, genetic evolution in populations, with a rather predictable process, development? Will any new insights emerge when we consider simultaneously the mechanisms operating at the population level and those acting within individuals? Patterns in the currently available data suggest that genetic evolution is, to some extent, predictable and that this predictability emerges from the interplay of developmental mechanisms with the mechanisms operating within populations. It seems to me that the current data are insufficient to rigorously test this hypothesis. But relevant new data are emerging rapidly, and they will allow robust tests of the combined roles of evolution and development in generating patterns of biological diversity.

This book is written for a reader, perhaps an advanced undergraduate or a beginning graduate student, who may have little or no formal training in either evolutionary biology or developmental biology. I assume that the reader has a basic understanding of undergraduate biology, including a basic understanding of DNA, genes, and proteins. Gene names are italicized and gene products are capitalized and set in roman type. I explicitly indicate whether I am referring to a gene (the DNA sequence in the genome: "the *Frigida* gene"), a gene product (either the RNA or protein encoded by the gene: "the

Frigida gene product"), or the protein product of the gene ("the Frigida protein"). I avoid abbreviations and acronyms (with a few exceptions: DNA, RNA) in the hope of clearing away trivial roadblocks to understanding.

Chapter 1 sets out the essential questions to be addressed, and Chapter 2 explores how long-term evolutionary patterns can be explained by principles from population genetics. Each of Chapters 3 through 5 starts out by exploring a different basic genetic concept that is fundamental both to developmental biology and to population genetics. I explore how these concepts allow us to draw connections between the disciplines and how a developmental perspective and a population-genetics perspective can be mutually informative. In Chapter 6, I briefly explore the profound consequences of two seemingly banal facts. First, all populations contain a finite number of individuals; and, second, in many species, individuals tend to stay in the same general locality generation after generation. In Chapter 7, I challenge the reader to think about development in a new way, backwards, because I think this approach to development clarifies why some genes contribute disproportionately to evolution. Finally, in Chapter 8, I review observations that suggest that genetic evolution is predictable and I discuss reasons why genetic evolution might be predictable.

While the early chapters provide brief sketches of some elementary concepts from evolutionary genetics and developmental biology, it is not possible to provide a complete introduction to either discipline in a book of this length. Thus, at the end of the book, I provide a list of recommended readings that expand on topics covered in each chapter. More advanced readers may find some of the early chapters elementary, but, even in these chapters, I have attempted to illustrate connections between evolution and development that have not been discussed widely in the literature. In addition to citations, the Notes at the end of the book provide expanded discussion of some topics mentioned only briefly in the main text.

# ONE

# IN ORDER TO FORM
# A MORE PERFECT
# UNION

Without [the light of evolution, biology] becomes a
pile of sundry facts some of them interesting or
curious but making no meaningful picture as a whole.

—Theodosius Dobzhansky, "Nothing in biology makes
sense except in the light of evolution"

D o frogs grow legs because thyroid hormone induces growth of legs or because legs evolved as an adaptation to life on land? Clearly, both explanations are true. The developmental answer addresses *how* legs grow. The evolutionary answer addresses *why* frogs evolved to grow legs. These two explanations do not compete to explain the existence of frogs' legs. Both explanations enrich our understanding of biology.

We can generalize from frogs' legs and observe that, at a mechanistic level, most biological diversity results from changes of gene function. In contrast, at an evolutionary level, selection on heritable variation causes biological diversity. These different levels of analysis provide complementary explanations for biological diversity. Failure to distinguish between levels of

1

analysis can lead to unproductive debates: "Frogs grow legs because of thyroid hormone"; "No, frogs' legs evolved as an adaptation to life on land." Usually, it is best to avoid comingling explanations at different levels of analysis.

In recent years, however, data has appeared that does not make sense except when viewed simultaneously from multiple levels of analysis. Here is one example.

In geographic areas with short or harsh summers, mouse-ear cress (*Arabidopsis thaliana*), a small, weedy annual plant, normally germinates in the fall, overwinters, and then flowers in the late spring. These plants require a period of cold to induce flowering, a process called vernalization. However, plants from some populations germinate in the spring and flower without requiring vernalization. These are called "rapid cyclers." Some rapid cyclers live in areas with long summers. Since the plants do not require vernalization, they can complete more than one life cycle in each growing season. Other rapid cyclers live in areas where harsh winters prevent autumnal germination and survival of seedlings.

In some *Arabidopsis thaliana* populations, mutations that incapacitate a single gene called *Frigida* cause plants to lose their ability to vernalize. These "null" mutations prevent production of a functional Frigida protein. These mutations delete all or part of the gene or introduce stop codons that truncate the protein prematurely. Different null mutations of the *Frigida* gene have arisen and spread at least twenty times and contribute to much of the variation in flowering time in natural populations of *Arabidopsis thaliana*.

Production of a functional Frigida protein causes plants to require vernalization. It therefore makes sense, from a developmental perspective, that mutations in the *Frigida* gene cause flowering time variation in natural populations. The null mutations in the *Frigida* gene can be seen as elegant paths to rapidly evolve a faster flowering time.

From an evolutionary perspective, however, the fact that flowering time has evolved by null alleles is compelling evidence that something out of the ordinary has occurred. Plant species closely related to *Arabidopsis thaliana* possess the *Frigida* gene, and even some distantly related plant species contain genes similar to *Frigida*. It is therefore unlikely that null mutations in the *Frigida* gene have contributed much to differences between closely related species; we would not find *Frigida* genes in these species if null mutations conferred a long-term evolutionary advantage. Null mutations appear to have spread in small populations that adapted rapidly to local conditions. The null mutations reduce flowering time, but, since these mutations have not spread widely to other *Arabidopsis thaliana* populations, they are likely to confer disadvantages on plants growing in other environments. This may be an example of short-term evolution leading a gene down an evolutionary dead end.

The natural history of *Arabidopsis thaliana* supports this view. *Arabidopsis thaliana* is native to Europe and to much of Asia, where it grows in recently disturbed ground. It has spread throughout the world, following in the footsteps of human agriculture. The plant reproduces primarily through self-fertilization, so even a single seed can found a new population of genetically similar individuals. Null mutations in the *Frigida* gene, which appear to provide a large advantage in some environments, may spread rapidly in these subpopulations in response to strong selection for rapid cycling.

While an understanding of the developmental role of the Frigida protein clarifies why mutations in the *Frigida* gene

3

generate changes in flowering time, population biology provides a context for understanding why these particular mutations have been selected.

These data contain a second surprising fact. In a mutagenesis experiment to identify genes involved in controlling vernalization, only three out of about 50 mutations occurred in the *Frigida* gene. The remaining mutations occurred in fifteen other genes. In all, mutations in at least 80 genes can affect flowering time. That is, mutations occurring randomly in the genome that affected vernalization would tend not to hit the *Frigida* gene. Why, then, does the *Frigida* gene harbor most of the evolved variation? Is there something special about the *Frigida* gene that makes it a favored target of natural selection?

The questions that arise by considering this example highlight the issues I will address in this book. Development teaches us which genes are available to generate particular phenotypic changes. Evolution teaches us what kinds of mutations may be favored over different time scales.

## SUMMARY

Each level of analysis warrants independent investigation by biologists. But new data do not make sense unless considered simultaneously from multiple levels of analysis. Interpretation of these new data requires a synthesis of evolutionary biology—particularly the biology of populations and of closely related species—and development biology.

TWO

# SCALE MATTERS

What we see depends mainly on what we look for.

—John Lubbock

There are multiple ways of looking at the same evolutionary pattern. We might focus on the broad sweep of evolutionary transitions amongst major groups of organisms or at the details of evolutionary change in populations. These different scales of inquiry allow us to ask different types of questions about evolution. But all evolutionary phenomena are rooted in a fundamental truth: evolution begins as variation within populations. We can therefore harmonize evolutionary observations at multiple scales by asking how they might be explained by population processes.

The evolutionary history of a group of organisms can be represented as a series of branching events, a phylogeny or phylogenetic tree. In the phylogeny in Figure 2.1, time is represented along the horizontal axis, moving forward from left to right. In this representation of a phylogenetic tree, the tree sits on its side. Living species sit at the tips of the tree on the right. The leftmost point, the root of the tree, represents the common ancestor of all the living species shown.

At this scale, we treat species as units. We talk about species evolving and splitting to generate new species. This is shorthand for a more complicated process. A species is a collection

5

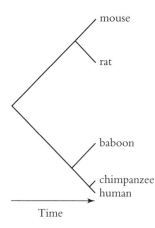

mouse

rat

baboon

chimpanzee
human

Time

FIGURE 2.1. This diagram represents a phylogeny of some vertebrate species over roughly the past 80 million years. The relative timing of speciation events is shown by the splitting of lines at different points on the time scale.

of individuals. The individuals reproduce and, over time, individuals in later generations may look different from their ancestors. Some of the individuals in a population may become reproductively isolated from other individuals in that population. This is shown as lines splitting in Figure 2.1. Often this splitting happens through geographic isolation. Two populations can then evolve independently from one another.

We can zoom in on one branch of the phylogeny to observe these events in more detail. To zoom in, we must be more specific about what we are observing. First, we will zoom in on individuals, as shown in Figure 2.2. We see that any individual has a complex ancestry. The clean lines of descent implied by the species phylogeny are absent at the individual level.

If we zoom in even further, however, to the gene level, the view changes. As illustrated in Figure 2.3, we again see clear lines of descent. Some gene lineages split and lead to multiple descendant lineages. Other lineages go extinct. Given enough time, all individuals in a population will carry copies of a gene, an allele, descended from a single individual. For example, the allele connected by the darker lines in the middle panel of Figure 2.4, has increased in frequency over time. This allele may

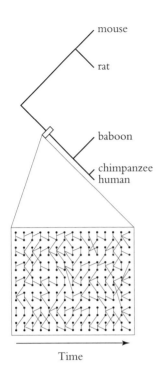

FIGURE 2.2. The dots and lines within the square represent hypothetical hereditary relationships between some individuals of the lineage ancestral to primates. Individuals are represented as dots, and each generation occupies a separate column. Two individuals (dots) mate to produce some offspring, or none. We can trace the ancestry of each individual back in time, but many individuals leave no descendants. Individuals that appear to have no ancestors migrated into this population, and their parents are not shown.

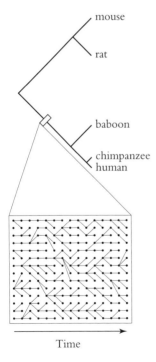

FIGURE 2.3. The dots and lines within the square now represent the hereditary relationships between alleles in some individuals from a population. Only one allele is shown for each individual. In diploid organisms, individuals carry two copies of most genes. A diploid individual may thus carry two different alleles of a gene.

7

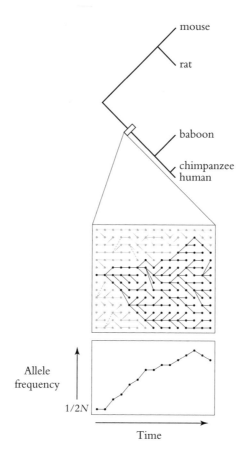

FIGURE 2.4. At any one time, all alleles can trace their ancestry back to a single copy some time in the past. If we follow the frequency of this allele in a population of diploid organisms, as in the bottom panel, we see that the frequency of this autosomal allele started at $1/2N$ and increases over time. Here, $N$ equals the number of individuals in the population.

ultimately become fixed in the population, it may *substitute* for the alternative alleles. What started as a single allele in a single individual may end up as a substitution in the population. We can also represent this process as a change in allele frequency over time, as shown at the bottom of Figure 2.4.

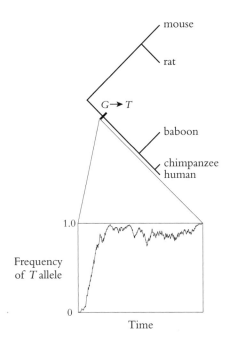

FIGURE 2.5. Replacement of a *G* allele with a *T* allele illustrated with a computer simulation of a new allele that eventually is fixed in a population of 1000 individuals. In this case, the new allele has a selective advantage of 2% over the old allele and shows complete dominance. The rugged line in the box shows the frequency of the *T* allele over time.

Changes in allele frequency represent the fundamental units of evolution. While we can observe phenotypic differences between individuals from different species and can reconstruct an evolutionary history as a phylogeny, the entire process has a mechanistic basis grounded in changes in allele frequency.

For example, evolutionary changes in DNA sequences often are represented by a mark or a symbol on a phylogenetic tree. Figure 2.5 shows a change in a DNA sequence represented as a change from guanine (*G*) to thymine (*T*) along one branch of a tree in the common ancestor of primates. This apparently single event represents two distinct processes. First, one chromosome in one individual in a population mutated from a $G$ to a $T$ at this particular position. Then, over multiple generations, the frequency of the $T$ allele increased until, eventually, it replaced the $G$ allele. The mutation became a substitution, as shown at the bottom of Figure 2.5.

Thus, the same event—substitution of a single allele—can be represented and studied at different scales. We can represent it as a mark on a phylogeny or as a change in allele frequency over multiple generations. Accounts of evolutionary events at large and fine scales sound different, but they refer to the same thing.

Another way to describe the process shown in Figure 2.5 is to say that the rodent sequences are different from the primate sequences because a substitution occurred in one position in the DNA sequence in the common ancestor of the primates. This is the kind of description used often in comparative developmental biology. This substitution may cause a phenotypic difference between species. However, an explanation at this level does not tell us anything about the specific population processes that caused the substitution. All we know is that the mutation became fixed in a population.

While descriptions of patterns above the population level may seem handy, often they hide useful details. Here is an example: Many proteins involved in the development of the fruit fly have amino acid sequences similar to proteins found in humans. For example, in Figure 2.6, I have aligned two regions of the protein products of the *Pax6* and *eyeless* genes that share similar amino acid sequences. These similarities result from the fact that the two proteins are related by evolutionary descent. More than 500 million years ago, one animal lived that carried

## Paired domain region

```
human Pax6    HSGVNQLGGVFVNGRPLPDSTRQKIVELAHSGARPCDISRILQTHAD
mouse Pax6    HSGVNQLGGVFVNGRPLPDSTRQKIVELAHSGARPCDISRILQTHAD
 fly Eyeless  HSGVNQLGGVFVGGRPLPDSTRQKIVELAHSGARPCDISRILQ----

human Pax6    AKVQVLDNQNVSNGCVSKILGRYYETGSIRPRAIGGSKPRVATPEVV
mouse Pax6    AKVQVLDNENVSNGCVSKILGRYYETGSIRPRAIGGSKPRVATPEVV
 fly Eyeless  ----------VSNGCVSKILGRYYETGSIRPRAIGGSKPRVATAEVV

human Pax6    SKIAQYKRECPSIFAWEIRDRLLSEGVCTNDNIPSVSSINRVLRNLA
mouse Pax6    SKIAQYKRECPSIFAWEIRDRLLSEGVCTNDNIPSVSSINRVLRNLA
 fly Eyeless  SKISQYKRECPSIFAWEIRDRLLQENVCTNDNIPSVSSINRVLRNLA
```

## Homeodomain region

```
human Pax6    GENTNSISSNGEDSDEAQMRLQLKRKLQRNRTSFTQEQIEALEKEFE
mouse Pax6    GENTNSISSNGEDSDEAQMRLQLKRKLQRNRTSFTQEQIEALEKEFE
 fly Eyeless  GENSNGGASNIGNTEDDQARLILKRKLQRNRTSFTNDQIDSLEKEFE

human Pax6    RTHYPDVFARERLAAKIDLPEARIQVWFSNRRAKWRREEKLRNQRR
mouse Pax6    RTHYPDVFARERLAAKIDLPEARIQVWFSNRRAKWRREEKLRNQRR
 fly Eyeless  RTHYPDVFARERLAGKIGLPEARIQVWFSNRRAKWRREEKLRNQRR
```

FIGURE 2.6. The protein products of the *Pax6/eyeless* genes of human, mouse, and fruit fly show significant similarity in amino acid sequence in two regions of the protein, the paired domain and the homeodomain. The relative locations of the paired domain and homeodomain are shown along the top, with the amino terminus (N) to the left. Alignment of the protein regions, including the paired domains and the homeodomains from all three species, are shown below. In this and later figures, each letter represents a different amino acid, according to the IUPAC code. Amino acids that are identical in all three genes are shaded and, with the exception of the deletion in the fly protein, those positions that differ in the fly are shown in bold. Over the entire protein, the human and mouse sequences differ at only one in 436 sites. Outside of these two regions, the two vertebrate sequences do not share extensive similarity with the fly sequence.

a gene similar to *Pax6*. Speaking loosely, one of its offspring went left and some of its descendants evolved into flies, and one went right and some of its descendants evolved into humans. More precisely, one copy of the gene ended up in the lineage leading to flies and one copy ended up in the lineage leading to humans. The two genes *Pax6* and *eyeless* are homologous. The sequences of homologous genes are rarely identical, residue by residue, along the entire protein, but they share more similarity than one would expect if amino acids were scattered randomly in proteins.

Many homologous genes can be found in distantly related species. When observed at this scale, the data can tell us a story about no more than the similarities and differences between species. When we think about the problem at a lower scale, at the level of populations, we gain insight into evolutionary mechanisms. Homologous proteins might share the same amino acid sequence for the trivial reason that the two species are very closely related and no mutations occurred that altered the amino acid sequence in the ancestors of these species. But we know that many new mutations are introduced into populations in every generation. It has been estimated, for example, that, in the human population, every possible mutation in the genome that is compatible with life has occurred about 240 times in just the last generation. Therefore, homologous genes are unlikely to result from an absence of mutations.

Homologous genes result from the failure of most mutations to substitute in populations. The vast majority of mutations are quickly lost. To explore this problem, we can classify mutations by their effects on fitness, which we can measure roughly as the number of surviving offspring. Organisms carrying a neutral mutation experience no change in their fitness on average. Therefore, the frequency of a neutral allele fluctuates randomly, and neutral mutations are lost often from populations. These random fluctuations are called genetic drift and can be observed as the squiggles of the line in the bottom part of Figure 2.5. (We will explore the causes and some of the consequences of genetic drift in later chapters.) Organisms carrying a deleterious mutation leave fewer surviving offspring, on average, than they would without the mutation, which increases the chances that the mutation will be lost from the population. Deleterious mutations experience what is called purifying selection. Finally, organisms carrying a beneficial mutation leave more surviving offspring, on average, than they would without the mutation, but even these beneficial mutations may still be lost due to genetic drift. I will discuss this last scenario in more detail in Chapter 3.

While individual neutral and selectively advantageous mutations may be lost from populations by genetic drift, the same mutations have probably arisen repeatedly during the history of a species. There are thus many chances for advantageous and neutral mutations to be substituted. Substituted mutations result in sequence differences between species. Therefore, the primary reason that homologous genes have *similar* DNA or amino acid sequences is that most mutations in these sequences reduced fitness and were eliminated from populations by purifying selection.

How do these population-genetic considerations illuminate the similarities between the homologous genes *Pax6* and *eyeless*? If all possible mutations in a gene were neutral, and if all DNA sites mutated at the same rate, then mutations in all sites would have substituted at the same rate. We can therefore get a general idea of how strongly purifying selection has acted by comparing the number of substitutions that have occurred at different locations in the genome. In the protein-coding region of genes,

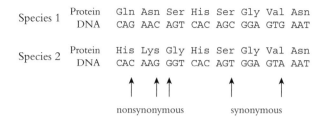

FIGURE 2.7. Nonsynonymous mutations alter the amino acid sequence of a protein, while synonymous mutations do not. Amino acids are represented by their three-letter codes, and nucleotides in triplet codons by their one-letter codes (A, G, C, and T). Arrows point to nonsynonymous and synonymous nucleotide replacements.

nonsynonymous mutations alter the amino acid sequence, while synonymous mutations do not, as shown in Figure 2.7. In most cases, synonymous mutations have weak effects or no effects on fitness. To a rough approximation, they are selectively neutral. If nonsynonymous mutations were also neutral, then they should be substituted at the same rate as synonymous mutations. However, nonsynonymous substitutions occur at a much lower rate than do synonymous substitutions. Between human and mouse genes, for example, the ratio of nonsynonymous to synonymous substitutions is about 0.15. This implies that at least 85% of nonsynonymous mutations that arose on these two lineages were deleterious and were eliminated by purifying selection. Purifying selection eliminated most mutations that altered the protein sequence. This is an explanation, at a population-genetic level, for why homologous genes can be recognized, even after a billion years of evolution.

One striking fact about homologous developmental genes is that they often perform similar functions in the two species. If we eliminate the gene in flies called *eyeless*, flies develop with much smaller eyes, as shown in Figure 2.8. That is, the *eyeless*

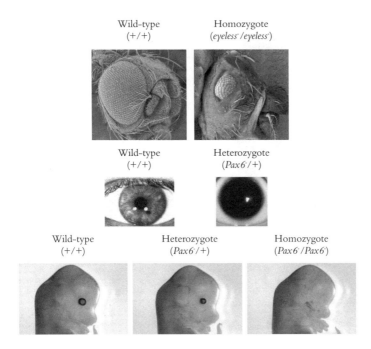

FIGURE 2.8. Mutations in homologous genes in fly, human, and mouse cause similar developmental defects. *Top row*: Scanning electron micrograph of a normal fly eye, on the left, and of the eye of an *eyeless* mutant, on the right. *Middle row*: Photograph of a normal human eye, on the left, and the eye of an aniridia patient, on the right. Aniridia results from a mutation in *Pax6*. *Bottom row*: Photographs of mouse embryos carrying two normal copies of *Pax6*, on the left, one defective copy of *Pax6*, in the center, and two defective copies of *Pax6*, on the right.

gene is required for normal development of fly eyes. The human homologue of the *eyeless* gene, called *Pax6*, is required for normal development of human eyes. Figure 2.8 shows that the same is true in mice. So these homologous genes appear to play similar developmental roles in flies and in mammals: both genes regulate eye development. This is true even though the fly has a so-called compound eye, constructed from hundreds of individual ommatidia that cannot be focused, while the mammalian eye is

FIGURE 2.9. Expression of the mouse Pax6 protein in the antenna of *Drosophila melanogaster* causes extra eyes to form on the antenna (*white arrow*).

a simple eye, with a focusing lens. At a coarse level, the homologous genes appear to have similar roles in development—both genes direct eye development—but, at a finer level, the two genes contribute to development of vastly different eyes.

Now, if the mouse Pax6 protein is expressed in flies, it promotes the development of eyes—fly eyes, of course, as shown in Figure 2.9. That is, the mouse Pax6 protein can substitute for the molecular function of the fly Eyeless protein in flies. To understand how this can possibly be true, we must delve further into the details of gene function.

Both the Pax6 and Eyeless proteins are transcription factors. I will discuss gene structure and transcription-factor function in more detail later, but for now note simply that transcription factors bind to specific DNA sequences to regulate gene transcription. This molecular function has been conserved in the *eyeless* and *Pax6* genes since the last common ancestor of flies and mammals. The conservation of molecular function in this case most likely derives from conservation of the amino acid sequence. This conserved amino acid sequence causes the proteins to adopt a similar three-dimensional shape. This particular shape allows the proteins to bind to specific DNA sequences and to activate transcription. Therefore, the mouse Pax-6 protein and the fly Eyeless protein bind to the same DNA sequences and both can regulate the activity of the correct genes in flies to produce fly eyes.

The Pax-6 and Eyeless proteins share similar amino acid sequences mainly in the DNA-binding domains of the proteins, as was shown in Figure 2.6. We can infer from the fact that these regions show conservation between flies and humans that these specific amino acids must have been subject to purifying selection. But purifying selection upon what? Selection acts by favoring or eliminating individuals with particular phenotypic traits—bigger or smaller eyes, for example. Thus, the conserved amino acid sequence must have been crucial to determining the same phenotypic characteristics reliably every generation since the common ancestor of flies and humans lived. What is this conserved aspect of the phenotype? What part of fly eyes resembles human eyes and has not changed since the ancestors of humans and flies diverged more than 500 million years ago?

The key to understanding this paradox is to recognize precisely what the conserved protein sequences, the DNA-binding domains, do. A transcription factor, such as the *Pax6* protein, binds to many DNA regions scattered throughout the genome to regulate dozens, hundreds, or thousands of genes. Let's imagine a single transcription factor that regulates just six genes, genes *a* through *f*.

Transcription factor

*a    b    c    d    e    f*

This transcription factor normally binds to all of these genes. We can imagine, however, that natural selection might favor a mutation in one of these target genes (*e*) that leads to loss of binding of this transcription factor.

Transcription factor

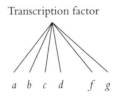

*a*  *b*  *c*  *d*     *f*

This change in regulation might increase fitness. It might, for example, generate a slightly larger eye. But the transcription factor still has to bind to all of the other genes that it regulates to produce a functional eye. One could also imagine a new mutation in the *g* gene that causes the transcription factor to bind to the *g* gene.

Transcription factor

*a*  *b*  *c*  *d*     *f*  *g*

The regulation of the *g* gene might also increase fitness. It might generate a slightly narrowed eye.

The *Pax6* gene, therefore, experiences purifying selection because it binds to a large battery of genes, all of which must be correctly regulated to produce an eye. The phenotypic features that result from development can evolve, nonetheless, by gain and loss of target genes. This explains why the Eyeless and Pax6 proteins can contribute to development of fly and mouse eyes, respectively, and why the mouse Pax6 protein can make fly eyes in flies. These homologous transcription factors must regulate a largely different set of downstream genes in flies and mice, but, at the molecular level, they have consistently performed the same job for 500 million years. They bind to similar DNA sequences of genes involved in eye development. Purifying selection has acted on the DNA-binding domains of these transcription factors. It is therefore likely that the common ancestor of flies and humans expressed a protein similar to the Pax6 and Eyeless proteins in the head to promote development of an eye, though the eye probably resembled neither a fly eye nor a human eye.

We have gone from an observation of pattern—conservation of parts of protein sequences between distantly related species—to an explanation of process. The explanation of evolutionary process was generated only by considering both population-level forces and molecular mechanisms. We could not have understood why the DNA-binding domain of a transcription factor was conserved unless we knew that transcription factors bind to many target sites in the genome. We were prompted to think in this way only once we realized that protein conservation between distantly related species demands an explanation, since population-level processes would not generate this pattern without purifying selection.

I have started with this simple example because it illustrates the dual role of population genetics and development in understanding a pattern in nature. Most biologists probably jumped to this conclusion without forcing themselves to think through the individual logical steps. Making these steps explicit will allow us to explore more complicated scenarios.

## SUMMARY

When we fail to account for the scale of inquiry, efforts to synthesize developmental and evolutionary biology are thwarted. Differences between species originated as variants in populations and cannot be understood fully without a firm grasp of population processes. Synthesizing developmental biology and evolutionary biology requires, in part, marrying studies at large and fine scales of evolution in the context of the mechanisms of development.

# THREE

# DOMINANCE

... those characters which are transmitted entire, or almost unchanged in the hybridization ... are termed the *dominant*, and those which become latent in the process *recessive*. The expression *recessive* has been chosen because the characters thereby designated withdraw or entirely disappear in the hybrids, but nevertheless reappear unchanged in their progeny ....

—Gregor Mendel, "Versuche über Plflanzen-hybriden (Experiments in Plant Hybridization)", as translated by William Bateson.

In diploid organisms (most animals, plants, and fungi), each individual possesses two copies of most genes, one on each homologous chromosome. Often, these two copies are not identical. They represent different alleles. Sometimes, the effect of one allele "hides" the effect of the other allele. This allele is termed dominant, the other is called recessive. The final phenotype depends on the state of both alleles in an individual, called the genotype. Dominance provides some initial clues to the role of development in evolution. To see these connections, we must first recognize that dominance can be defined in ways that seem superficially different but that are, in fact, similar at a deeper level. First, we can focus on the phenotype—the appearance, or physiological state, or behavior of an individual—and label alleles as dominant or recessive, based on the phenotypic effects produced by different combinations of alleles. Inferences based on population genetics, however, require estimates of fitness. We can then explore the effects of dominance on fitness. I will first explain the phenotypic definition and then discuss the fitness definition.

Geneticists define an allele as dominant when organisms containing one or two copies of the allele present the same altered phenotype. In the top graph of Figure 3.1, the *A* allele is dominant to the *a* allele for some phenotypic feature, let's say pigmentation. Animals with the genotype *aa* display the recessive phenotype: white fur. Animals with either genotype *Aa* or *AA* display the dominant phenotype: black fur. An animal carrying at least one *A* allele produces black fur.

Dominant alleles are typically capitalized, but this is simply a naming convention that helps geneticists to keep track of the dominance relationships of alleles. We could, in principle, define the *A* allele as recessive, as in the middle graph of Figure 3.1. In this case, the heterozygote *Aa* produces white fur and the homozygote *AA* produces black fur. Upon discovering this dominance relationship, most geneticists would simply switch the capitalization of the alleles, and this may therefore seem like a trivial exercise. A more important point is that there is nothing about dark pigmentation, per se, that makes alleles that cause dark fur dominant. As we will see, dominance emerges from the specific role of a gene in development and in how mutations alter gene function. The middle graph of Figure 3.1 also helps to illustrate an important evolutionary principle. If a population is full of *a* alleles, the evolutionary fate of the new *A* allele depends on whether the *A* allele is dominant or recessive. If the *A* allele is recessive, then it is effectively "hidden" in the

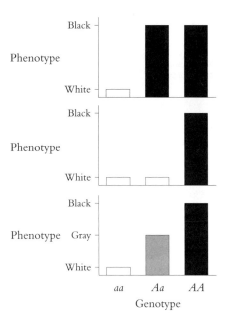

FIGURE 3.1. The phenotypes of animals with different genotypes can vary, depending on the effects of dominance. *Top graph:* Allele *A* is dominant to allele *a* for coat color. Animals carrying two copies of the recessive allele, *a*, have a white coat. Animals carrying at least one copy of the dominant allele, *A*, have a black coat. *Middle graph:* Allele *A* is recessive to allele *a* for coat color. Animals carrying at least one copy of the *a* allele have a white coat. Animals carrying two copies of the recessive allele, *A*, have a black coat. *Bottom graph:* Allele A shows incomplete dominance for dark coat color. The amount of pigmentation is quantitatively related to the number of *A* alleles an animal carries.

population until two *Aa* heterozygotes mate. Then, some of the offspring may be *AA* and will produce black fur.

Many alleles also show incomplete dominance, as illustrated in the bottom graph of Figure 3.1. In this case, the *aa* homozygote produces white fur and the *AA* homozygote produces black fur. The *Aa* heterozygote produces an intermediate phenotype, say gray fur. In this case, if the *A* allele arises as a new

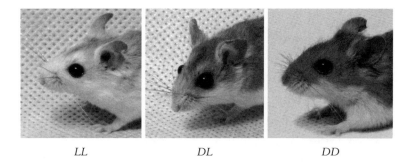

LL                    DL                    DD

FIGURE 3.2. This figure illustrates incomplete dominance of naturally occurring alleles of the *Agouti signaling protein* gene in the oldfield mouse, *Peromyscus polionotus*. Photographs of mice with three genotypes are shown: on the left, a homozygote for the *Light* allele = *LL*; in the center, a heterozygote for the *Light* and *Dark* alleles = *LD*; and on the right, a homozygote for the *Dark* allele = *DD*.

mutation in a population, it can be detected immediately because it produces a new phenotype, but not the same phenotype that is produced by the homozygote *AA*. Figure 3.2 shows an example of incomplete dominance for naturally occurring alleles in a population of beach mice.

Geneticists long ago noticed that the majority of mutations that disrupt gene function tend to be recessive. This is probably because only specific kinds of mutations can cause phenotypic dominance. The majority of mutations that disrupt gene activity cause a quantitative reduction in the amount of gene product produced. These quantitative changes are unlikely to cause phenotypic dominance for the following reason. Many proteins act as enzymes in pathways. Even transcription factors can be considered enzymes; they act in pathways leading to the synthesis of specific RNAs. When enzymes act in pathways, changes in protein activity tend to be buffered by the activity of other enzymes in the pathway. Thus, the flux through a path-

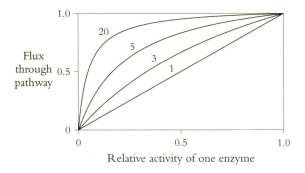

FIGURE 3.3. Flux through an enzymatic pathway is relatively insensitive to changes in enzyme activity. Flux through a hypothetical pathway of 1, 3, 5, or 20 enzymes, each of which has an equivalent effect on flux, is shown as a function of the activity of a single enzyme in the pathway.

way tends to resemble the curves shown in Figure 3.3. When multiple enzymes participate in a biochemical process, the flux curves show a pronounced plateau at levels of high enzyme activity.

Most enzymes are produced in sufficient quantity that their activity places them high on the plateau of flux through the pathway, toward the right of Figure 3.3. In this case, reducing activity of a single enzyme by 10%, or 20%, or 30% from the normal level has little effect on flux. Flux through the pathway continues to sit high on the plateau. Even reducing activity by one-half, which occurs in a heterozygote carrying one null allele, causes a relatively small change in flux. For example, *Drosophila melanogaster* flies carrying only one functional copy of an *Alcohol dehydrogenase* gene do not experience a significant reduction in the oxidation rate of ethanol, as can be seen by comparing *F/N* with *F/F* and *S/N* with *S/S* in Figure 3.4. Figure 3.4 also illustrates that natural variation in Alcohol dehydrogenase protein activity causes only small changes in

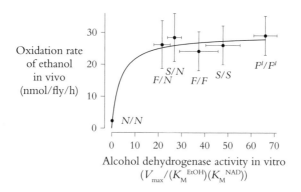

FIGURE 3.4. Variation in activity of enzymes encoded by naturally occurring *Alcohol dehydrogenase* alleles from *Drosophila melanogaster* produce little difference in the oxidation rate of ethanol in vivo, even when combined with a null allele. The figure shows values for six genotypes: homozygotes for three wild functional alleles, Fast$^d$ = $F^d/F^d$, Fast = $F/F$, Slow = $S/S$; a homozygote for a null allele = $N/N$; and heterozygotes between two of the functional alleles and the null allele, $F/N$ and $S/N$. Curve for 20 enzymes in a pathway, each with equal effects on flux, is fitted for illustrative purposes.

ethanol metabolism. Large changes in flux, the kinds of changes that might alter the final phenotype, typically require a large decrease in the activity of a single enzyme. This is a biochemical explanation for why most null mutations have recessive phenotypic effects. The reduction in enzyme activity caused by one copy of a null allele produces only a small change in the organism's final phenotype. Many proteins behave as expected under this theoretical model and this is the likely explanation for the predominance of recessive mutations.

While changes in catalytic activity are unlikely to generate phenotypic dominance, changes in the pattern of gene expression can easily cause phenotypic dominance. For example, a dominant allele of the *Antennapedia* gene causes expression of Antennapedia protein in an organ, the antenna, where it is not

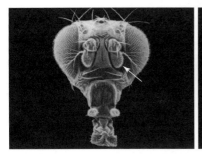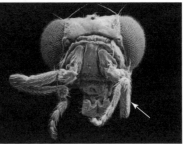

FIGURE 3.5. Frontal view of the head of *Drosophila melanogaster* is shown on the left. One antenna is indicated with a white arrow. On the right, the head of a fly carrying a dominant allele of the *Antennapedia* gene produces legs in place of antennae.

normally expressed. This causes the antenna to adopt the morphology of a leg, as shown in Figure 3.5.

Mutations causing altered gene regulation, leading to dominant phenotypes, have occurred also during evolution. For example, in many organisms, the eye-lens crystallins, which are proteins that maintain transparency and the correct light diffraction in the lens of the eye, are enzymes, such as Lactate dehydrogenase protein and Argininosuccinase lyase protein. These enzymes have evolved to be expressed at high levels in the lens. These high levels of expression resulted from mutations that increased expression specifically in the lens, without any changes in the catalytic activity of the enzymes.

Qualitative changes in protein function also can cause phenotypic dominance. If an enzyme evolves a qualitatively new catalytic activity, the quantity of enzyme activity for this new function increases from zero to some positive value. As shown in Figure 3.3, this would cause a large change in flux. It is difficult for an existing gene to evolve a qualitatively new catalytic function, since a gene is normally constrained to continue performing its current function. However, new genes arise frequently by mutations that duplicate existing genes. At first, the duplicate gene usually contains the same DNA sequence as the original gene. Over time, the two copies of the gene will substitute mostly different mutations. In many cases, since only one copy of the gene is necessary, one copy is quickly lost as a result of mutations that generate a null allele. Null mutations arise frequently, since there are so many ways to inactivate a gene, and these mutations are often effectively neutral, since the second copy of the gene is still present. Occasionally, however, one of the duplicated copies evolves a new function, whilst the other copy retains its original biochemical role.

Gene duplication, therefore, initiates a lot of evolutionary novelty at the molecular level. For example, while some eye-lens crystallins still retain their ancestral enzyme functions and have not resulted from gene duplication, in many cases eye-lens crystallins have evolved from enzymes after gene duplication. In a dramatic example, some fish that live in subzero waters have evolved antifreeze proteins that capture ice in the bloodstream to prevent ice crystals from puncturing their cells. This antifreeze protein evolved from a duplicated pancreatic *trypsinogen* gene. Trypsinogen protein is secreted from the pancreas into the small intestine, where it is activated to form trypsin. The antifreeze protein no longer exhibits trypsin activity and it has evolved extensive changes that allow it to bind to ice. In addition, the protein is expressed in the liver to allow its secretion into the bloodstream. Thus, this evolutionary innovation required changes both in protein structure and in expression pattern. Both changes were likely to have caused dominant phenotypic effects.

Changes in the catalytic activity of the ice-binding protein occurred through changes in the amino-acid sequence. But how do changes in gene expression pattern evolve? To answer

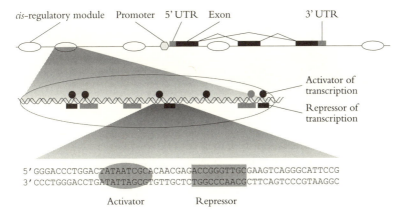

FIGURE 3.6. This figure illustrates the structure of a generic eukaryotic gene. The exons are indicated in gray, connected by diagonal lines that indicate how the exons are spliced to form the mature mRNA. The DNA sequences between exons of a single gene are called introns. Often, alternative mRNAs can be generated by the splicing together of alternative exons, as indicated by the alternative diagonal lines connecting exons. This results in proteins containing different amino acid sequences. The dark gray parts of the exons encode the protein and the light grey parts encode the 5' and 3' untranslated regions of the mRNA. The untranslated regions (UTRs) form part of the mature mRNA and are required for proper translation. The promoter is located directly 5' (upstream) of the 5' UTR and contains DNA sequences required for proper localization of the basal transcription apparatus to the gene. The ovals indicate *cis*-regulatory modules, which may be located 5', 3', or within the introns of genes. One *cis*-regulatory module is magnified to illustrate the locations and DNA sequences bound by individual transcription factors. For purposes of illustration, activators are placed above the DNA strand and repressors below.

this question, we must first explore the structure of genes in more detail.

As shown in Figure 3.6, all genes are composed of a string of DNA that encodes two broadly different kinds of information. First, the DNA encodes the gene product in regions called exons. In most cases, the exons are transcribed into mRNA that is later translated into proteins. Exons encode the polypeptide chains of proteins with a remarkably orderly system of sequential triplets of DNA bases.

Second, the DNA encodes instructions for when and where genes will be transcribed. Unlike exons, these instructions are not organized into tidy regions. Instead, we say that the instructions are found in the *cis*-regulatory DNA. At first glance, the instructions in *cis*-regulatory DNA seem to be scattered, almost willy-nilly, in the general region of—but normally upstream of—the exons. There does seem to be some structure to the distribution of these instructions; instructions for one job tend to be close to one another. To see the order, though, and perhaps why it evolved, it is helpful to focus first on the individual pieces of these instructions.

The individual units making up the *cis*-regulatory DNA of a gene consist of short DNA sequences, or motifs, of about 6 to 20 base pairs per unit. Each motif can be bound by a specific transcription factor. Of course, transcription factors can bind to DNA motifs only when they already are expressed in cells; so, most of the time, in any particular cell, most of the transcriptional instructions are not bound by any transcription factor. When a transcription factor does bind to a particular site, however, it then either promotes recruitment of, or retards recruitment of, the basal transcription apparatus to the promoter. The basal transcription apparatus is a large conglomeration of proteins that transcribes mRNA. Many transcription factors can act as repressors or as activators depending on physical interactions with other proteins.

Most transcription factors bind only weakly to specific DNA sequences; they display low binding affinity. Small changes in the DNA sequence, like the mutation of a single base, can dramatically alter the binding affinity. Small changes in the sub-cellular environment also can alter the binding affinity. Transcription factors bind not only to DNA, but they bind also to other transcription factors with low affinity. Thus, transcription factors sometimes recruit other transcription factors to the same local DNA region. These weak binding interactions between transcription factors can greatly increase the binding affinity of transcription factors for DNA. Clustering of motifs in a short stretch of a DNA sequence can therefore catalyze cooperative binding of transcription factors. This cooperative binding probably increases the accuracy and robustness of gene regulation.

Sometimes, physically contiguous clusters of transcription-factor binding sites encode a spatially and temporally discrete transcriptional output. We sometimes refer to these as *cis*-regulatory modules or enhancers. For example, the *even-skipped* gene, which encodes a transcription factor, is expressed in seven stripes in the early *Drosophila melanogaster* embryo, as shown in Figure 3.7. These stripes of expression help to establish the segmented body of the fly. Most stripes are driven by independent *cis*-regulatory modules. The *cis*-regulatory region driving expression of the second most anterior stripe (stripe 2) has been studied in considerable detail.

The 400 base-pair region of DNA that drives expression of the *even-skipped* gene in stripe 2 contains multiple binding sites for the transcription factors encoded by four genes. The combined activity of these four transcription factors defines the spatial pattern of expression in the following way. Two transcription factors, the Bicoid protein and the Hunchback protein, promote transcription from the *even-skipped* gene. As shown in the drawings in Figure 3.7, the Bicoid protein is present at high levels at the anterior end of the embryo, and it grades to lower levels in

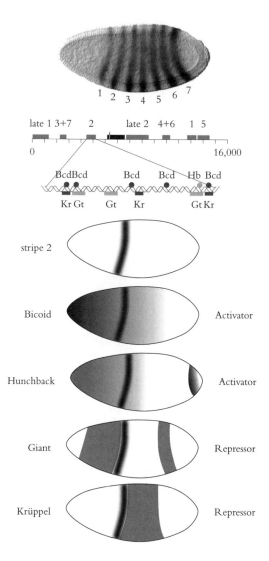

FIGURE 3.7. Expression of the *even-skipped* protein, shown at the top, is
driven by at least six *cis*-regulatory enhancers spread over 16,000 base
pairs of the *even-skipped* gene, shown below the embryo. The exons
are shown in black. Some stripes, such as stripe 2, are driven by independent enhancers. Some require activity of two different *cis*-regula-
tory regions at different times. Stripes 3 & 7 and 4 & 6 appear to be
driven by single enhancers. These apparently overlapping functions of
single enhancers may reflect overlap of the required binding sites, or

*(continued on facing page)*

the posterior. The Hunchback protein is expressed at high levels in the anterior half of the embryo. If the stripe 2 enhancer were to contain binding sites only for the Bicoid and Hunchback proteins, then these two proteins would drive expression of the Even-skipped gene product throughout the anterior half of the embryo. The precise spatial expression pattern of the Even-skipped gene product results from repression by two other transcription factors, the Giant protein and the Krüppel protein. In the anterior half of the embryo, the Giant protein is expressed with sharp anterior and posterior boundaries. The posterior boundary of Giant protein expression abuts the anterior boundary of Even-skipped protein expression in stripe 2. The Giant protein represses transcription from the *even-skipped* gene by binding to the stripe 2 enhancer in anterior cells. Repression by the Krüppel protein, which is expressed in a band in the middle of the embryo, similarly defines the posterior boundary of Even-skipped protein expression in stripe 2.

It is easy to see how changes in gene expression might occur as a result of changes in the DNA of the *even-skipped* gene stripe 2 enhancer. Loss or weakening of binding sites for the Giant protein would shift the anterior boundary of Even-skipped gene product expression in stripe 2 to the anterior. In

---

simply that there are two enhancers located close together that have not yet been distinguished. The stripe 2 enhancer contains multiple binding sites for two activators and two repressors. The expression pattern driven by the stripe 2 enhancer alone and the expression patterns of the two activators—Bicoid and Hunchback proteins—and the two repressors—Giant and Krüppel proteins—are shown below. The combined activity of these four transcriptional regulators defines stripe 2, but not other stripes. The other *cis*-regulatory enhancers may be driven by different combinations of these and other transcriptional activators and repressors.

later chapters, I will review examples of novel evolutionary changes in gene-expression patterns caused by changes in the transcription-factor binding sites of *cis*-regulatory modules.

We expect that *cis*-regulatory mutations causing gene expression in a new domain of a developing organism will be at least semidominant if not fully dominant. This follows from the fact that half a dose of an enzyme usually produces much more than half of the activity provided by a full dose. For the same reason, *cis*-regulatory mutations that reduce or eliminate gene expression will tend to be recessive.

The genetic definition of dominance leads biologists inward, toward the cellular and subcellular mechanisms that generate the organism. To explore the role of dominance and other developmental phenomena in evolution, we must exploit definitions that point outwards, from the individual to the population level. We must explore how dominance affects reproductive fitness. The absolute fitness of individuals, which we might define as the total number of surviving offspring, is not the most convenient way to think about fitness for evolutionary studies. Instead, we will consider the effects of alleles on relative fitness—the reproductive success of an individual in comparison with other individuals in the population. It is convenient to define the average of the fitnesses of all individuals in the population as 1. We then compare the fitness of an individual carrying a new mutation with this population average fitness. For example, imagine that all individuals in the population carry the *aa* genotype at one particular gene; the average fitness of individuals with the *aa* genotype is then equal to the population average fitness, or 1. A new mutation arises at this gene, generating allele *A*. The individual carrying this new mutation has the genotype *Aa* and may experience fitness equal to, less than, or greater than the fitness of the currently widespread genotype. We will find it convenient to define the fitness of a homozygote for this new mutation, *AA*, as $1 + s$, where $s$ is called the selection coefficient. Thus, neutral mutations have a selection coefficient of 0; deleterious mutations have a negative selection coefficient; and adaptive mutations have a positive selection coefficient. This quantification of fitness allows population geneticists to use the tools of mathematics to answer evolutionary questions.

We can now explore the effects of dominance on evolutionary dynamics. Figure 3.8 illustrates the effects of dominant and recessive alleles on fitness. An individual heterozygous for a dominant allele has the same fitness as an individual homozygous for a dominant allele, as shown in the top graph of Figure 3.8. A recessive allele, allele *A* in the middle graph of Figure 3.8, changes fitness only when homozygous. Some alleles show incomplete dominance, where the heterozygote enjoys a fitness intermediate between that of the two homozygotes. The degree of dominance is often indicated with the symbol *h*. In the bottom graph of Figure 3.8, I illustrate the special case of semidominance, where the heterozygote enjoys a fitness exactly intermediate between the two homozygotes: $h = 0.5$. I have not illustrated several other important cases, namely when the heterozygote experiences higher or lower fitness than both of the homozygotes.

Figure 3.8—showing dominance for fitness—appears similar to Figure 3.1—showing dominance for phenotype—only because I decreed, arbitrarily, that dark phenotypes should take higher values than light phenotypes in Figure 3.1. Without

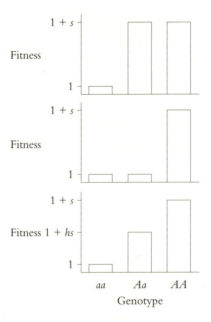

FIGURE 3.8. The effects of different genotypes on fitness depend on dominance. In the top graph, the $A$ allele is dominant to the $a$ allele. Fitness is relative and refers to a comparison of genotypes. In this example, $s$ is the selection coefficient in favor of the $Aa$ and $AA$ genotypes. In the middle graph, the $A$ allele is recessive to the $a$ allele. In the bottom graph, the $A$ allele shows partial dominance over the $a$ allele. The $Aa$ heterozygote has average fitness of $1 + hs$, where $h$ represents the degree of dominance of the $A$ allele. In this case $h = 0.5$.

measures of fitness, we have no basis for saying that one pheno-type is superior to another. Even for a single pair of alleles, the phenotype and fitness can, at least in principle, display different degrees of dominance. We cannot simply assume that fitness will show the same pattern of dominance as the phenotype.

Populations of stickleback fish provide a striking example of how phenotypic dominance can be an uncertain guide to fit-ness dominance. Marine stickleback populations, consisting predominantly of a heavily armored "complete" morph pos-

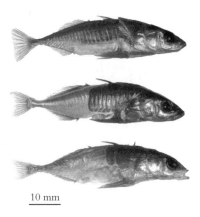

10 mm

FIGURE 3.9. From top to bottom, the "complete," "partial," and "low" morphs of the stickleback *Gasterosteus aculeatus*.

sessing multiple bony plates (shown in the top of Figure 3.9) have repeatedly invaded new freshwater habitats. Once in these freshwater habitats, the "complete" morphs tend to decrease in frequency, and the weakly armored "low" morph possessing few bony plates (shown at the bottom of Figure 3.9) increases in frequency. The difference in armor between morphs is caused by variation at multiple genes, but variation at one gene, called *Ectodysplasin*, has a strong effect on the number of bony plates. Homozygotes for the "complete" *Ectodysplasin* allele display more bony plates than heterozygotes, so this allele shows intermediate phenotypic dominance. We can imagine a simple explanation for the loss of bony plates in freshwater populations: reduced predation in freshwater habitats combined with the metabolic cost of producing bony plates causes selection against individuals carrying "complete" *Ectodysplasin* alleles in freshwater. An experimental study illustrates, however, that this simple model provides, at best, an incomplete explanation for the loss of heavily armored morphs. In this experiment, fish carrying the "complete" and the "low" alleles of *Ectodysplasin*

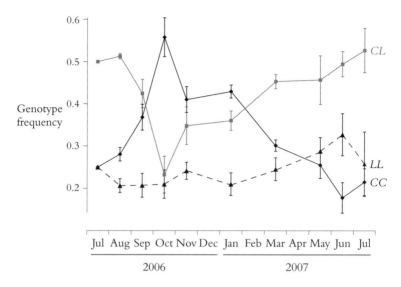

FIGURE 3.10. Frequency of three *Ectodysplasin* genotypes varies over time after introduction of fish into experimental ponds. Data represent means and standard errors for four replicate ponds. The fate of the homozygotes for the "complete" allele (*CC*) is shown in the solid black line; the fate of the homozygotes for the "low" allele (*LL*) is shown in the dashed black line; and the fate of the heterozygotes for the "complete" and "low" alleles (*CL*) is shown in the solid gray line.

were introduced into artificial freshwater ponds, and the frequencies of the three possible *Ectodysplasin* genotypes were tracked over the course of one year. As shown in Figure 3.10, early in the life of these fish, from July to October of 2006, heterozygotes carrying one allele each of the "low" and "complete" alleles of *Ectodysplasin* suffered a strong fitness decrement. Later in life, from November 2006 to May 2007, these heterozygous fish enjoyed much higher fitness than did the two homozygotes. Although selection favored the loss of bony plates overall in this experiment, phenotypic dominance displayed a poor correlation with fitness dominance across the year, as shown in Figure 3.11.

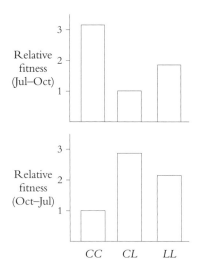

FIGURE 3.11. Dominance for fitness in stickleback fish in the experiment illustrated in Figure 3.10 varies depending on when fitness is measured. Dominance for the phenotype, shown in Figure 3.9, is therefore a poor predictor of dominance for fitness in this case.

Given these complications, one might ask why we would bother to study dominance for fitness. I will review two important reasons. First, fitness dominance influences the fate of deleterious mutations. Second, dominance can influence the fate of advantageous mutations.

When new deleterious mutations are introduced into a population, selection will act to remove them. Deleterious mutations with precisely recessive effects will not affect fitness when found in heterozygotes. They reduce fitness only when organisms are homozygous for the allele, and homozygotes are produced only when heterozygotes mate, which happens infrequently when alleles are rare. In comparison, a new deleterious mutation with even weak dominance will reduce fitness, on average, in all individuals unfortunate enough to carry the allele. Thus, selection can act more efficiently against mutations with partial dominance than against completely recessive mutations. At the same time that purifying selection is eliminating these deleterious mutations from populations, new deleterious mutations are constantly being reintroduced. This implies that purifying selection

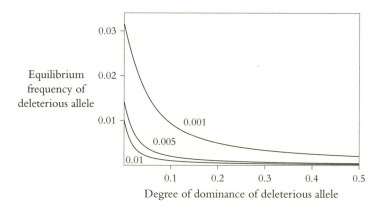

Degree of dominance of deleterious allele

FIGURE 3.12. The degree of dominance has a large effect on the equilibrium frequency of a deleterious allele under mutation-selection balance. The three curves illustrate the equilibrium frequencies for alleles with selection coefficients against them of 0.001, 0.005, and 0.01. With low levels of dominance, these deleterious alleles can reach appreciable frequency. With higher levels of dominance, deleterious alleles are kept at much lower allele frequencies. For this illustration, I assumed a mutation rate of $10^{-6}$ for each class of deleterious mutations.

will never succeed in eliminating all deleterious mutations from a population and, indeed, all populations harbor many deleterious mutations at low frequency. The rate of introduction of new deleterious mutations eventually will equal the rate of their selective elimination, attaining what is called mutation–selection balance. Figure 3.12 shows that dominance has a significant effect on the equilibrium frequency of deleterious mutations under mutation–selection balance. Even small increases in dominance above zero greatly reduce the equilibrium frequency of deleterious mutations.

To see how dominance can affect the fate of advantageous mutations, we first must reconsider the process of allelic substitution discussed in Chapter 2. A new allele starts as a single copy and may, eventually, become fixed—substituted—in the population. However, most new adaptive alleles do not substitute. When a new allele arises by mutation, the allele is present initially as a single copy. This simple fact has some surprising consequences. An allele present as a single copy in a population will usually go extinct very quickly as a result of genetic drift. Imagine that we represent the gametes from a population containing 50 diploid individuals as a bag full of 100 marbles. 99 marbles are white and one, our new allele, is black. Now we wish to sample the next generation by selecting 100 marbles randomly; we are simulating the random sampling that causes genetic drift. We select marbles without looking in the bag. We replace each marble after we check its color. Sometimes, we will sample a new population with exactly one black marble, and sometimes we will sample a population containing more than one black marble. About 37% of the time, however, we will fail to select the rare black marble, and the new population will consist solely of white marbles. (This is the probability of selecting a white marble 100 times $= 0.99^{100}$.) Thirty-seven percent of the time we will fail to select this single black marble in the first generation. Our random sampling of marbles simulates two ever-present factors in nature that generate genetic drift: the lottery of meiosis and random variation in fecundity. Near the boundary of a frequency of zero, random sampling essentially sucks new mutations into oblivion at a high rate.

We can use our marbles to simulate natural selection by biasing our sampling to select black marbles, as natural selection might favor a particular phenotype. If our black marble is slightly larger than each white marble, then it will tend to rise to the top of a well-mixed bag of marbles, thus increasing the chances that we will select the black marble. However, even in this case, we will often sample a new generation without a black marble. The boundary effect has again sucked the black marble into oblivion. In nature, alleles at low frequency have a high probability of being lost by drift, and this effect can overwhelm any fitness advantage provided by the new allele. When there are few copies of anything, no matter how precious (alleles, keys, coins, ancient manuscripts), accidental loss of a few copies results in the complete elimination of all copies.

Nonetheless, the larger size of the black marble does favor the selection of black marbles for the next generation. In nature, too, selection has a strong effect. The probability that a selectively advantageous semidominant mutation is ultimately fixed is equal to approximately twice the selection coefficient enjoyed by the heterozygote (approximately $2hs$). Imagine that a mutation arises in a population of 50,000 diploid individuals and confers a 1% fitness advantage on the heterozygote ($aa = 1, Aa = 1.01, AA = 1.02$); one percent of the time, an individual carrying this allele produces an extra surviving offspring. The probability of fixation of this selectively advantageous mutation is about 0.02. In contrast, a new neutral allele has a probability of

$$\frac{1}{2N}$$

of ultimately being fixed, or, in this case, 0.00001. In comparison with a neutral allele, a selective advantage of only one percent increases the probability of fixation in this population by a factor of two thousand.

New mutations that generate dominant fitness improvements can be favored by natural selection as soon as they arise. This does not guarantee, of course, that the allele will be fixed.

Genetic drift ensures that, most of the time, most new mutations, even many of those that confer dominant fitness advantages, will be lost. Nonetheless, natural selection is more likely to fix alleles with dominant fitness effects than alleles with recessive fitness effects. A new recessive allele must first spread to intermediate frequency, entirely by chance, so that two heterozygotes can mate and produce offspring homozygous for the recessive allele. Thus, recessive alleles spend more time under the influence of purely random events than do dominant alleles and recessive alleles are therefore more likely to disappear without a trace.

Mutations with semidominant fitness effects can be detected in heterozygotes, but dominant alleles would increase fitness in heterozygotes even more. Semidominant alleles are therefore less likely to become fixed than are dominant alleles, but they are more likely to become fixed than are recessive alleles. The effect of dominance on the probability of fixation is illustrated with a computer simulation in Figure 3.13. The stronger the dominance of an allele, the more likely that it will be fixed by natural selection. Over most of the range of dominance, the probability of fixation of a new mutation equals approximately $2hs$. This effect of fitness dominance is caused by the battle between natural selection and genetic drift that occurs soon after the mutation first appears.

The effect of dominance on the probability of fixation of new mutations was discovered by J. B. S. Haldane and is called "Haldane's sieve" in his honor. It counts, I believe, as the first theoretical insight into how developmental mechanisms can influence evolutionary processes. To the extent that phenotypic dominance reflects fitness dominance, alleles displaying stronger dominance are more likely to be fixed. We can imagine many

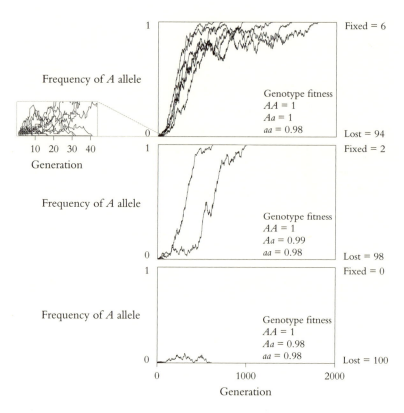

FIGURE 3.13. Each panel illustrates simulations of 100 diploid populations, each with a population size of 1000 individuals. The new
adaptive allele $A$ started off as a single copy—that is, in a heterozygote—and the simulations were run until all populations were fixed
for one of the alleles. More populations fixed strictly dominant adaptive alleles, shown in the top panel, than fixed semidominant alleles,
shown in the middle panel. None of the populations fixed for the
adaptive recessive alleles, as shown in the bottom panel. In most populations, the new adaptive allele was lost in the first few generations,
which is illustrated with an expanded view, to the left, of the first
forty-four generations of the simulations with the dominant allele.

FIGURE 3.14. Melanic (left) and typical (right) morphs of the peppered moth, *Biston betularia*.

simple ways in which phenotypic dominance may be related to fitness dominance. For example, mutations that alter pigmentation can generate dominant, semidominant, or recessive phenotypic effects. If the fitness of an organism depends on the degree of color matching between pigmentation and background, then pigmentation may influence fitness. This would create a correlation between phenotypic dominance and fitness dominance. For example, peppered moths, *Biston betularia*, come in multiple color morphs, two of which are shown in Figure 3.14. The alleles conferring the dark morph are dominant to the alleles conferring the light morph. After coal-fired industry darkened the trees in England in the early twentieth century, dark morphs increased in frequency, perhaps because the dark pigmentation provided better camouflage. In some heavily polluted regions, the dark morph increased in frequency to nearly 100%. Following a reduction in the use of coal, in the late twentieth century, the light morph increased in frequency, reducing the frequency of the dark morph to near 0%. During the decades of heavy industrialization, the heterozygote dark morph appears to have enjoyed fitness similar to that of the homozygote; following the cleanup of industrial pollution, both dark morphs suffered a similar fitness decrement. Thus, in this case, phenotypic dominance appears to closely reflect fitness dominance.

At first sight, it might seem that Haldane's sieve predicts that adaptation should occur most frequently by fixation of dominant alleles. That prediction is unlikely to be valid under most conditions, however, for at least two reasons. First, the prediction assumes that mutations with dominant and recessive fitness effects occur with equal frequency, and we don't know currently if this is true. As mentioned earlier, the majority of mutations studied by developmental biologists are phenotypically recessive, although these mutations may affect fitness in heterozygotes. If recessive advantageous mutations occur more frequently than do dominant advantageous mutations, then we may not detect an excess of dominant mutations contributing to adaptation.

Second, Haldane's sieve relies, in part, upon the assumption that all adaptive alleles are generated by new mutations. This might be a sensible assumption, if all populations simply evolved to maximize fitness in a single, stable environment. But there is ample evidence that environmental conditions vary and that individuals emigrate to new environments. In a variable environment, neutral or deleterious alleles that were already present in a population may suddenly become advantageous. One source of potentially advantageous alleles is the pool of deleterious mutations that are maintained by mutation–selection balance. If these previously deleterious alleles become advantageous in a new environment, then Haldane's sieve may fail to act. Under a wide range of assumptions, the probability of fixation for preexisting alleles maintained by mutation–selection balance is independent of dominance. The reason is that, while alleles showing stronger dominance are more likely to fix, there are fewer such alleles in a population because they are kept at lower frequency under mutation–selection balance. These effects tend to cancel each other out. This scenario is not as far-fetched as it might sound. For example, some insects carrying mutations conferring insecticide resistance suffer reduced fitness in the absence of the insecticide.

While dominance begins to reveal connections between development and population genetics, we cannot make simple predictions about the overall frequency with which dominant alleles will contribute to adaptation. Selection following a recent change in environmental conditions may make use of variation maintained under mutation–selection balance. In this case, selection will not generate a strong bias for mutations showing stronger dominance. If the environment subsequently remains stable for many generations after the environmental change, then adaptation in the long term will make increasing use of alleles introduced by mutation. In such a case, alleles showing stronger dominance would be more likely to fix. We can, therefore, compose a new prediction about the role of Haldane's sieve in adaptation. Given multiple mutations that can generate the same phenotypic outcome, adaptation resulting from fixation of new mutations is expected to more often utilize mutations that show stronger dominance.

## SUMMARY

A dominant allele masks the effects of an alternative allele at a locus. Phenotypic dominance results from the action of multiple gene products in enzymatic pathways. Halving the dose of a gene product rarely generates half the phenotypic effect. Qualitative changes in the catalytic activity of a gene product and changes in gene expression can cause dominant phenotypic effects. Changes in gene expression can occur through changes in the *cis*-regulatory regions of genes.

We can explore the role of dominance in evolution by defining dominance for fitness. Fitness dominance influences the frequency of deleterious mutations maintained by mutation–selection balance. New, selectively advantageous alleles displaying stronger dominance have a higher likelihood of substituting. However, changes in the environment may cause selection for previously neutral alleles or for previously deleterious alleles maintained under mutation–selection balance. Dominance has little effect on the probability of fixation of mutations maintained under mutation–selection balance. Thus, selection immediately following changes in the environment may select mutations with a broad range of dominance effects, whereas long-term selection in a stable environment may select a biased subset of mutations displaying stronger dominance.

FOUR

# PLEIOTROPY

Since the gene exists in every cell of the body,
it may be expected to affect the organism as a
whole, even if its most striking effect is on some
particular organ or function.

—J. B. S. Haldane, *The Causes of Evolution*

The available evidence indicates that pleiotropy is
virtually universal.

—Sewall Wright, *Evolution and the Genetics of Populations,
vol. 1 (Genetics and Biometric Foundations)*

Since the dawn of modern genetics, scientists have realized
that single genes often function in multiple tissues. For
example, the first mutation found in *Drosophila melanogaster*, a
mutation that inactivates the *white* gene, causes the fly's normally red eyes to be white, but it also removes pigmentation
from the male testis, changes the shape of the female spermatheca, and reduces lifespan and viability. Geneticists therefore
invented the term pleiotropy for mutations that cause phenotypic effects on two or more body parts that do not share a
clear functional connection with each other.

This definition of pleiotropy relies on a subjective evaluation
of whether phenotypic effects are related. If, for instance, a

mutation causes smaller body size, do we say that the mutation has a specific effect on total body size or that it has pleiotropic effects making all organs smaller? Further, pleiotropy can be confused with different physical manifestations of the same molecular cause. For example, in 1865, Gregor Mendel described a gene with two alleles that caused pea seeds to be wrinkled or smooth. But these alleles also alter the shape of the starch granules in the seeds, the sugar content of seeds, and the capacity of seeds to absorb water. In 1927, Theodosius Dobzhansky considered these to be pleiotropic effects of one gene. We now know, however, that this gene encodes a starch-branching enzyme and that the wrinkled allele is null, producing no active enzyme. Wrinkled peas thus contain lower levels of starch than do smooth peas, but they contain higher levels of free sugars. These free sugars generate a higher osmotic potential, causing wrinkled seeds to absorb more water than smooth seeds. The wrinkled seeds lose a higher proportion of their water when they mature and, since the outer covering of the seeds does not shrink to the same extent as the cotyledons, the seeds wrinkle. Do these different phenotypic effects of this mutation represent pleiotropic effects or simply different descriptions of the same phenomenon? Calling these effects pleiotropic or saying that the gene itself is pleiotropic does not clarify matters.

The concept of pleiotropy takes on greater utility if we define pleiotropy in relative terms. One gene may contribute to the development of eyes only and another to the development of eyes and toes. The second gene has more roles in development than the first gene has and is therefore more pleiotropic.

The term pleiotropy becomes most useful, however, when applied to different mutations in the same gene. As early as

1920, geneticists recognized that different mutations at the same gene can affect subsets of organs. In 1922, Hermann Muller noted that different mutations of the *truncate* gene in *Drosophila melanogaster* caused shortened wings, eruption of the thorax, and lethality, or some combination of these phenotypic effects. We can delineate all the roles of a gene by examining the effects of a null allele. If a null mutation in a particular gene has more distinct effects than other mutations in the same gene, then the gene plays pleiotropic *roles* in development. A gene with pleiotropic roles participates in the development of multiple phenotypic characteristics. An individual mutation may affect some or all of these roles, in which case the mutation has pleiotropic *effects*. Some mutations in pleiotropic genes will affect only one role; they have specific effects.

The pleiotropic roles of genes can result either from a single protein performing multiple distinct molecular jobs or from expression of a protein in multiple tissues. I discuss each source of pleiotropy separately.

The first source of pleiotropy results from the interaction of a protein with several other molecules to perform a task. For example, transcription factors bind DNA, bind other transcription factors, and bind the basal transcription apparatus. Each function requires a different region of the protein. Mutations in these different regions often have similar phenotypic effects, such as abrogating transcription of all target genes. But, in some cases, mutations in different regions have more specific effects than do null mutations.

For example, the *Ultrabithorax* gene, which encodes a transcription factor, is required for normal development of many organs of the thorax and abdomen, including: the balancer organs called halteres on the third thoracic segment; the second

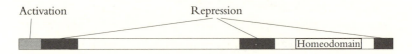

FIGURE 4.1. Separate regions of the Ultrabithorax protein mediate transcriptional activation and repression functions. At least three separate regions are required for full transcriptional repressive function. The DNA binding domain is labeled as Homeodomain.

and third pairs of legs; the shape, size, and bristle pattern of the third thoracic and first abdominal segment; and organs of the nervous system and the gut. The Ultrabithorax protein represses transcription of some genes in some tissues and activates transcription of some genes in others. These different repressive and activator functions involve different domains of the protein, as shown in Figure 4.1. A region near the start of the protein is required for the activation functions, and at least three separate regions are required for the full repressive function.

The separation of the activation and repression domains and the redundancy of the repression domains may allow mutations in a single domain to have relatively specific phenotypic effects. For example, Figure 4.2 shows that deletion of 24 amino acids from the end of the Ultrabithorax protein causes subtle phenotypic effects. Flies heterozygous for this deletion mutation show no detectable phenotypic changes. Flies homozygous for this deletion produce a few extra trichomes on the legs, display a slight increase in the size of the haltere, and sometimes produce a small bristle on the haltere.

These phenotypic effects are far more subtle than the effects of null mutations in the *Ultrabithorax* gene, which cause the entire third thoracic segment to develop like the second thoracic segment and which kill the animal during the larval stage. Flies

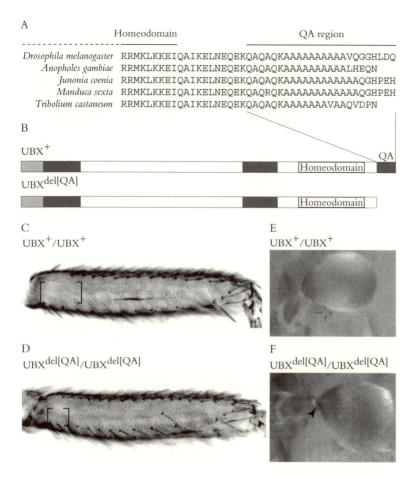

A

Homeodomain                                    QA region

Drosophila melanogaster  RRMKLKKEIQAIKELNEQEKQAQAQKAAAAAAAAAAVQGGHLDQ
Anopheles gambiae  RRMKLKKEIQAIKELNEQEKQAQAQKAAAAAAAAAALHEQN
Junonia coenia  RRMKLKKEIQAIKELNEQEKQAQAQKAAAAAAAAAAAAQGHPEH
Manduca sexta  RRMKLKKEIQAIKELNEQEKQAQRQKAAAAAAAAAAAAQGHPEH
Tribolium castaneum  RRMKLKKEIQAIKELNEQEKQAQAQKAAAAAAAVAAQVDPN

B

UBX$^+$

UBX$^{del[QA]}$

C
UBX$^+$/UBX$^+$

D
UBX$^{del[QA]}$/UBX$^{del[QA]}$

E
UBX$^+$/UBX$^+$

F
UBX$^{del[QA]}$/UBX$^{del[QA]}$

FIGURE 4.2. Deletion of a conserved 24 amino-acid motif from the end of the Ultrabithorax protein has minor effects on the adult phenotype. (A) This region contains a conserved QA-rich domain. The dashed line under the word homeodomain indicates that the homeodomain continues further to the left. (B) Deletion of the QA motif of the Ultrabithorax protein. (C) On the posterior second femur, wild-type Ultrabithorax protein represses trichome development on a patch of cuticle. (D) Legs produced by flies homozygous for the deletion mutation repress trichomes on a smaller patch of cuticle. (E) A haltere produced by a wild-type animal. (F) Flies homozygous for the deletion mutation produce a slightly enlarged haltere that sometimes carries an ectopic bristle (arrow).

FIGURE 4.3. A normal adult *Drosophila melanogaster* is shown on the left. On the right, a fly carrying multiple mutations that eliminate expression of Ultrabithorax protein from most of the third thoracic segment developed with a second pair of wings in place of halteres.

carrying mutations in the *Ultrabithorax* gene that almost entirely eliminate expression of Ultrabithorax protein from the third thoracic segment develop with wings in place of halteres, as shown in Figure 4.3. Although different domains of the Ultrabithorax protein are required for different functions, it is nonetheless unlikely that evolution of these domains has contributed much to phenotypic change amongst insects. This is because all of these domains are highly conserved across insects, as was shown in Figure 4.2A, indicating that these domains have experienced primarily purifying selection during insect evolution.

The second source of pleiotropic roles results from expression of the same protein in different tissues or from expression of the same protein at different times in the same tissue. These spatial and temporal patterns of differential expression usually are controlled by information encoded in the *cis*-regulatory regions of genes. Gene expression in each domain may be determined by a different *cis*-regulatory enhancer. Mutations may then alter these independent enhancers to cause specific phenotypic effects.

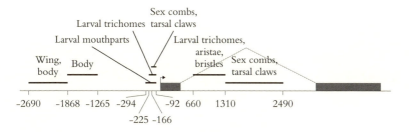

FIGURE 4.4. Largely separable *cis*-regulatory modules of the *yellow* gene drive Yellow protein expression in different tissues. The exons of the *yellow* gene are shown as gray boxes. The locations of enhancers are indicated with horizontal lines, and distances from the site of transcription initiation are indicated in base pairs.

For example, the Yellow protein, which is required to make darkly pigmented cuticle in *Drosophila melanogaster*, is expressed in many different body regions. Figure 4.4 illustrates that expression in different regions is determined by largely separable *cis*-regulatory modules of the *yellow* gene.

Novel patterns of Yellow protein expression have evolved by alterations in existing *cis*-regulatory modules and through origin of new *cis*-regulatory modules. The wings of most *Drosophila* species are translucent, but some species have evolved one or multiple spots of dark pigmentation. As shown in Figure 4.5, the expression pattern of Yellow protein correlates with dark pigmentation patterns in some species. Expression of Yellow protein in these spots is not sufficient to generate pigmentation, but high levels of Yellow protein may be required to produce the strong pigmentation in these spots.

Detailed studies of the *yellow* gene have revealed several ways that novel expression patterns can evolve through changes in *cis*-regulatory regions. *Drosophila biarmipes* evolved several new transcription-factor binding sites within a *cis*-regulatory mod-

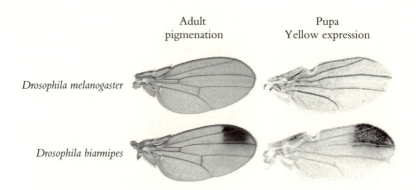

Adult
pigmenation

Pupa
Yellow expression

*Drosophila melanogaster*

*Drosophila biarmipes*

FIGURE 4.5 Expression of the Yellow protein in the developing wing, shown on the right, is correlated with the production of melanin in the adult wing, shown on the left.

ule that already drove expression of Yellow protein in the wing, as shown in Figure 4.6. At least two of these transcription-factor binding sites are required for repression of expression in the posterior of the wing, like this ⬤. At least one transcription-factor binding site is required for activation of expression in the distal portion of the wing, like this ⬤. The overlap of this repression and activation generates a spot of Yellow protein expression in the anterior, distal domain of the wing, like this ⬤. Thus, in this case, the novel expression pattern evolved through modifications of an existing enhancer that was already active in the wing.

A second example illustrates how a similar expression pattern resulted from evolution of an entirely new enhancer. Figure 4.7 shows that *Drosophila tristis*—independent of other species with wing spots—evolved a spot of Yellow protein expression in the wing. Expression of Yellow protein in a spot is driven, in this case, by a newly evolved enhancer located within the first intron.

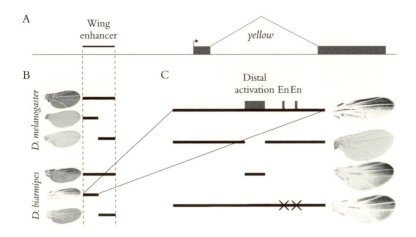

FIGURE 4.6. Evolution of a wing-spot enhancer in *Drosophila biarmipes* by mutations in a wing enhancer. (A) The *yellow* gene is shown with the location of the 5' wing enhancer originally identified in *Drosophila melanogaster*. (B) The *Drosophila melanogaster* wing enhancer is sufficient to drive expression of a reporter gene throughout the wing blade. Partial enhancers containing either the left or right side of the *Drosophila melanogaster* wing enhancer are insufficient to drive reporter expression in the wing blade. The homologous DNA region from *Drosophila biarmipes* drives reporter gene expression in *Drosophila melanogaster* throughout the wing blade and more strongly in an anterior (up) and distal (right) domain. A smaller region from the left of the *Drosophila biarmipes* wing enhancer drives expression in an anterior-distal spot. A smaller region from the right drives reporter gene expression throughout the wing blade. (C) The "spot" enhancer includes a domain required for activation in the distal domain and at least two binding sites for a repressor, the Engrailed protein, which is expressed in the posterior of the wing. Deletion of the distal activation region eliminates distal expression, and this small region on its own drives expression in the distal wing blade. Mutation of the two binding sites for the Engrailed protein reduces posterior repression.

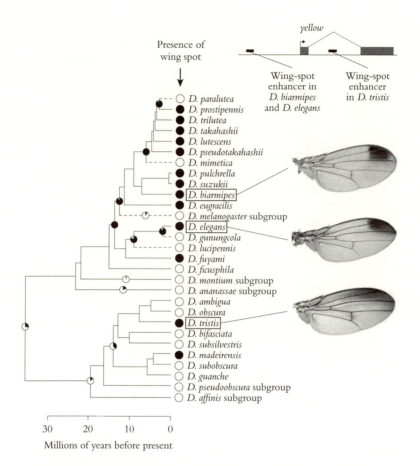

FIGURE 4.7. A novel wing-spot enhancer evolved in the intron of the *yellow* gene in *Drosophila tristis*. A phylogeny of species of the *Drosophila melanogaster* and *Drosophila obscura* species groups is shown on the left. Presence or absence of a wing spot is indicated for all species, and the probability of a wing spot in many of the common ancestors is shown as the percentage of black in each circle. *Drosophila tristis* appears to have evolved a wing spot independently of *Drosophila biarmipes* and *Drosophila elegans*. The locations of the *cis*-regulatory elements that drive expression of the Yellow gene product in wing spots for the three species are shown at top right, and images of male wings from each species are shown below right.

Have these enhancers evolved by minor changes, such as single nucleotide substitutions, or by larger changes, such as wholesale duplication of enhancers from other genomic regions? Unfortunately, these species are so distantly related that many nucleotide changes have accumulated in their genomes. These multiple nucleotide substitutions have obscured the individual nucleotide changes driving Yellow protein expression in the wings of *Drosophila biarmipes* and *Drosophila tristis*. However, other species provide some indication of how many nucleotide changes are required to generate similar changes in gene expression. Figure 4.7 shows that *Drosophila gununcola* and *Drosophila elegans* are closely related, and that *Drosophila gununcola* has lost the wing spot since diverging from a common ancestor that most likely had a wing spot. At least two and no more than seven point mutations generate this difference between the species. In this case, loss of Yellow protein expression in a wing spot was caused by multiple point mutations in an enhancer region.

Thus, pleiotropic roles of genes can evolve through the fixation of multiple mutations in *cis*-regulatory enhancers resulting in novel temporal and spatial patterns of gene expression.

At any particular time, populations may experience selection on multiple phenotypic characteristics simultaneously. Selection may act at the same time on parasite resistance, cold tolerance, finger length, brain size, and hairiness, to name just a few phenotypic characteristics. If mutations with pleiotropic effects could simultaneously solve several ecological problems, then they would be favored over mutations that ploddingly solved only one problem at a time. But there are probably few mutations that can solve several ecological challenges at once.

Here is one way to explore this issue. Start with a population that is well adapted to its environment; most mutations will be deleterious. Now imagine that the environment changes slightly. In the old environment, small wing sizes were optimal, but, in the new environment, intermediate wing sizes are optimal. Consider a gene with only one function, a nonpleiotropic gene that controls wing size. Some mutations in this nonpleiotropic gene can improve fitness by altering wing size. Which ones? As shown in Figure 4.8, we will assume that fitness is distributed symmetrically around the optimum wing size, so that wings slightly larger or slightly smaller than the optimum both confer an equal fitness advantage. Therefore, any nonpleiotropic mutations that move the wing size closer to the optimum—either undershooting, or overshooting, or jumping directly to the optimum—will increase the fitness of their bearer. But mutations that overshoot the optimum by too much—in this case, by more than twice the distance to the optimum—will decrease fitness. Of course, any mutations that make wings smaller will also be deleterious.

If a gene has pleiotropic roles, then we need to consider the possibility that mutations in this gene will have pleiotropic effects. Mutations may affect not only wing size but also other phenotypic traits, such as leg size and eye size. Many genes are known to contribute to development of multiple organs, and mutations in these genes may have pleiotropic effects. The smattering of fly species shown in Figure 4.9 occupy many locations in the three-dimensional space defined by leg, wing, and eye size, implying that many lineages have traversed this three-dimensional space over the past several hundred million

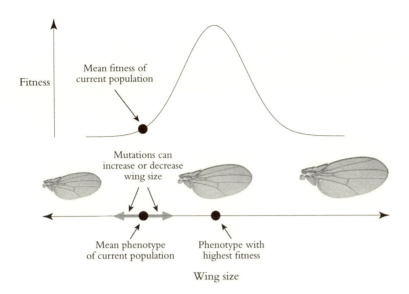

FIGURE 4.8. This figure illustrates a one-dimensional model of adaptation. The phenotype with the highest fitness is a wing of intermediate size, shown in the middle. Individuals in the population currently have wings smaller than the optimum. New mutations either can have no effect or can increase or decrease wing size. Mutations that decrease wing size are selectively disadvantageous. Mutations that increase wing size—in this case, to less than twice the distance to the optimum—increase fitness.

years. How do species evolve through this multidimensional space?

Consider a population, shown as a sphere in Figure 4.10, that currently sits some distance from the optimum, indicated by a star. To reach the optimum, or even to get close, the population must evolve along all three axes defining this multidimensional space. Some mutations may affect only one organ, such as mutations in a gene with pleiotropic roles that alter the activity of a single *cis*-regulatory enhancer. These mutations will tend to move the phenotype along only a single axis, such as wing

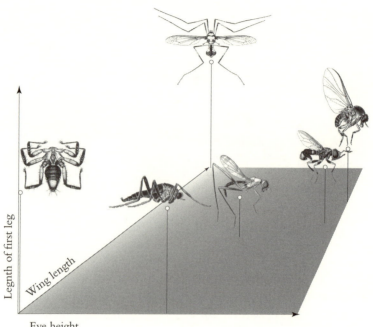

Length of first leg

Wing length

Eye height

FIGURE 4.9. This figure illustrates some fly species scattered in a three-dimensional space representing eye height, wing length and length of first leg relative to total body length. Fly drawings have been scaled to approximately the same overall size to allow comparison of relative organ sizes. The relative position of each species in this space is indicated by an open circle and a vertical line connected to a plane defined by the eye height and wing length axes. The two flies closest to the observer develop without wings.

length. But many mutations, such as many mutations that alter protein function, will simultaneously affect many traits. Pleiotropic mutations may move individuals in a random direction through this three-dimensional space, but most steps in a random direction will tend to move the phenotype further away from the optimum.

Of course, some genes contribute to development of more than three phenotypic features. Mutations in these genes may

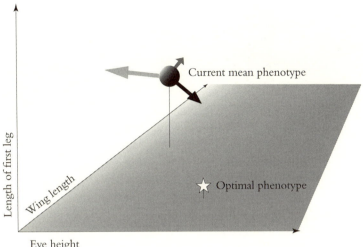

Length of first leg

Wing length

Eye height

FIGURE 4.10. This figure presents a multidimensional model of adaptation. In this example, the phenotype consists of three traits: eye height, wing length, and length of first leg. The location of the optimum phenotype in this three-dimensional space is indicated by a star. The population currently sits some distance from the optimum. This may occur, for example, because environmental change displaces the optimum. New mutations, represented as arrows emanating from the current mean phenotype, can move the phenotype in a random direction. Most new mutations will therefore move the population away from the optimum. As the number of dimensions increases, a smaller fraction of mutations are likely to move the population toward the optimum.

have highly pleiotropic effects. As the number of potential pleiotropic effects increases, it becomes increasingly unlikely that a random step in any direction will move the phenotype closer to the optimum. This assumption has been partially tested in the yeast *Saccharomyces cerevisae*. Four thousand seven hundred and eighteen strains of yeast were generated, each of which carried a deletion of a different single gene. Of these 4718 mutated strains, 1667 showed pleiotropic effects on various phenotypic characteristics. The relative fitness of each mutant strain was

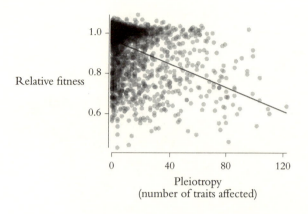

FIGURE 4.11. Deletion mutants of yeast display a negative correlation between pleiotropic effects and relative fitness. Each datum is light gray, and darker grey represents the piling up of multiple data points at the same location. The line represents the least-squares regression.

then compared with the original strain, and mutations with stronger pleiotropic effects were found to confer lower fitness than mutations with few pleiotropic effects, as shown in Figure 4.11. The theoretical considerations discussed above combined with this empirical evidence suggest that mutations with fewer pleiotropic effects are more likely to improve fitness than are mutations having extensive pleiotropic effects.

So far, we have considered only the probability that a single mutation will be beneficial. To consider advantageous mutations as a class, we must first remember that many beneficial mutations are lost from populations by genetic drift. Nonetheless, mutations conferring large fitness gains are more likely to substitute than those conferring small fitness gains, because the probability of fixation depends on the selection coefficient. Thus, when a population sits far from the fitness optimum, mutations of large effect are more likely to substitute than are mutations of small effect.

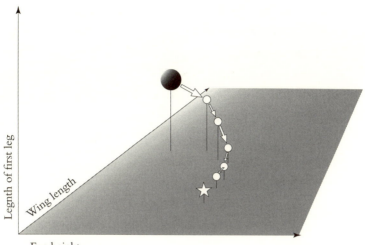

FIGURE 4.12. This figure illustrates evolution via an adaptive walk. The population was initially located at the dark sphere. The optimal phenotype is located at the star. The population evolved through the sequential substitution of multiple mutations that moved the phenotype closer to the optimum with each successive substitution. In this hypothetical example, most mutations that substituted had at least partially pleiotropic effects. The initial mutation, for example, reduced the length of the first leg and increased eye height, but did not alter wing length.

In nature, adaptation probably rarely involves substitution of only a single mutation. Instead, novel ecological demands, such as a change in temperature or the introduction of a new parasite, probably require multiple adjustments to the phenotype. We might visualize adaptation as a walk toward the fitness optimum, with multiple mutations substituting in the population over time, as illustrated in Figure 4.12. In support of this model, many studies have demonstrated that the individual phenotypic differences between species are usually caused by multiple mutations, usually found in multiple genes. As a population evolves closer to the optimum, mutations of smaller and smaller effect, on average, will tend to be substituted, since the distance to the optimum gets smaller with each successive substitution event. Thus, early in the adaptive walk, we expect that mutations of relatively large effect may substitute. But as the population evolves closer to the optimum, substitutions of mutations causing finer adjustments will predominate.

∽

How well do these theoretical considerations agree with empirical observations? We currently know few of the mutations causing phenotypic differences between species. However, the patterns observed in many of these examples are consistent with theoretical expectations. Mutations contributing to species differences tend to have highly specific (nonpleiotropic) effects. Here I review two examples.

In the nematode worm *Caenorhabditis elegans,* the *lin-48* gene, which encodes a transcription factor, is expressed in the hindgut, the excretory duct, the male tail, and in a variety of other cells. Null alleles of the *lin-48* gene cause pleiotropic effects, including causing the excretory duct to develop in a more anterior position than it normally would. The excretory duct in the nematode species *Caenorhabditis briggsae* does not express Lin-48 protein, and it develops in a more anterior position than in *Caenorhabditis elegans*, as shown in Figure 4.13.

Loss of Lin-48 protein expression in the excretory duct of *Caenorhabditis briggsae* was caused by at least four mutations in a *cis*-regulatory region of the *lin-48* gene. Each of the *cis*-regulatory mutations that arose in *Caenorhabditis briggsae* slightly reduced expression of Lin-48 protein in the excretory duct. These mutations do not alter expression levels of Lin-48 pro-

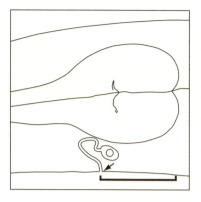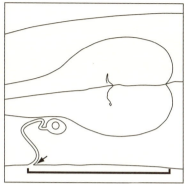

FIGURE 4.13. The position of the excretory duct has evolved between *Caenorhabditis* species. In *Caenorhabditis elegans*, on the left, the excretory duct (*arrow*) is close to the posterior of the pharynx (*distance marked with a bracket*). In *Caenorhabditis briggsae*, on the right, the excretory duct (*arrow*) is positioned in a more anterior position (*distance marked with a bracket*).

tein in other organs, such as the hindgut and male tail. These mutations have small and specific effects on Lin-48 protein expression in the excretory duct.

A second example reveals a similar pattern. Fruit fly larvae are decorated with a complex pattern of cuticular projections called microtrichia, or trichomes for short. Cells in different regions of the larva produce trichomes of different sizes and shapes. The large trichomes on the ventral surface of the larva aid in locomotion. In most *Drosophila* species, larvae produce a lawn of fine trichomes on their dorsal and lateral surfaces. For reasons that are not yet known, the pattern of these trichomes evolves rapidly between species, as shown in Figure 4.14. For example, the dorsal and lateral surfaces of first-instar larvae of *Drosophila melanogaster* and most closely related species are almost completely covered with trichomes. *Drosophila sechellia*, by contrast, produces only a few dorsal rows of trichomes.

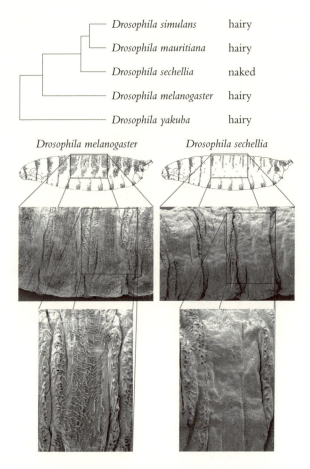

FIGURE 4.14. *Drosophila sechellia* first-instar larvae have evolved a naked dorsal and lateral cuticle. A phylogeny of the *Drosophila melanogaster* species subgroup, illustrated at the top, shows that the naked phenotype has evolved in the *Drosophila sechellia* lineage. Drawings and scanning electron micrographs illustrate the hairy phenotype of *Drosophila melanogaster* and the naked phenotype of *Drosophila sechellia*. No intraspecific polymorphism for the hairy and naked phenotype has yet been discovered.

This morphological innovation in *Drosophila sechellia* was caused by evolutionary changes at the *shavenbaby* gene. The *shavenbaby* gene, coincidentally, is the homologue of the *Caenorhabditis elegans lin-48* gene discussed above. Like the *lin-48* gene, the *shavenbaby* gene encodes a transcription factor.

The Shavenbaby protein acts like a genetic switch. Cells that accumulate Shavenbaby protein produce trichomes, and cells that do not accumulate Shavenbaby protein produce naked cuticle. During embryogenesis, Shavenbaby protein is expressed in a complex pattern that precisely prefigures the pattern of trichomes that will appear later on the larva. As illustrated in Figure 4.15, the complex expression pattern of the Shavenbaby gene product is determined by at least three enhancers—proximal, medial, and distal—found approximately 10,000, 28,000, and 50,000 base pairs upstream of the first exon of the *shavenbaby* gene, respectively. Each of these enhancers produces part of the complete expression pattern of the Shavenbaby protein. The proximal enhancer drives expression of Shavenbaby protein in ventral stripes and in some of the dorsal rows; the medial enhancer drives Shavenbaby protein expression in ventral rows and in wide swatches of dorsal rows; and the distal enhancer drives Shavenbaby protein expression mainly in lateral patches, and a bit in dorsal and ventral stripes.

The fact that each of the three enhancers drives Shavenbaby protein expression in both conserved and evolved domains of the shavenbaby expression pattern means that no single mutation could cause the observed evolutionary changes in morphology. A deletion that removed an entire enhancer, for example, might remove trichomes that are still present in *Drosophila sechellia*. In fact, each of the three enhancers has evolved to have

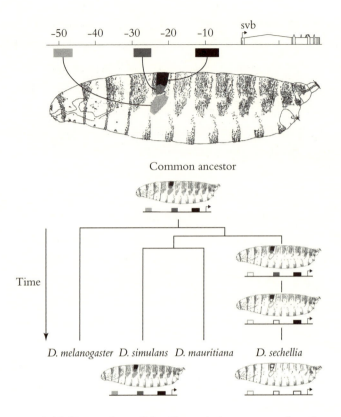

FIGURE 4.15. Expression of the Shavenbaby gene product is regulated by at least three distinct *cis*-regulatory regions. A map of the *shavenbaby* gene is shown at the top. Numbers are kilobase pairs from the transcription start site of the *shavenbaby* gene. Three DNA regions, marked by light, medium, and dark gray boxes drive expression in lateral, dorsolateral and dorsal regions of the domain that has evolved in *Drosophila sechellia*. In addition, all three enhancer regions drive expression in domains that differentiate stout trichomes that are conserved between species (not shown). The expression domains are indicated for only one segment in the larval drawing. At the bottom, a phylogeny illustrates evolution of all three enhancer regions, leading to the "naked" trichome pattern in *Drosophila sechellia*.

an altered function. All three enhancers still drive essentially the same ventral pattern of expression as well as expression in some of the dorsal rows, but each enhancer has lost its ability to drive expression of Shavenbaby protein in dorsal or lateral regions.

A larva homozygous for a null allele of the *shavenbaby* gene produces almost no trichomes. The evolved trichome pattern in *Drosophila sechellia* is a much less extreme change in the phenotype than the loss of trichomes caused by a null allele. Therefore, the *cis*-regulatory mutations at the *shavenbaby* gene that have caused phenotypic evolution are less pleiotropic than a null mutation of the *shavenbaby* gene. Each *cis*-regulatory mutation at the *shavenbaby* gene causes only part of the evolved trichome pattern in *Drosophila sechellia*.

*Cis*-regulatory mutations are also likely to be less pleiotropic than many coding changes that might alter Shavenbaby protein function. Mutations altering the Shavenbaby protein would likely affect development of all trichomes, not just trichomes in specific domains. In addition, the *shavenbaby* gene shares most of its exons with the *ovo* gene, a gene required for ovary development. Changes in exons would likely affect both trichome and ovary development. *Cis*-regulatory changes in the *shavenbaby* gene minimize pleiotropic effects associated with altering trichome patterns.

Many genes that participate in development, such as *lin-48* and *shavenbaby*, possess complex *cis*-regulatory regions and their gene products in turn regulate multiple other genes. This generates an asymmetry; multiple inputs regulate transcription of the *lin-48* and *shavenbaby* genes, and their gene products then regulate transcription of multiple other genes. Mutations in *cis*-regulatory regions can alter how the *lin-48* and *shavenbaby* genes are regulated by only one or some of the multiple inputs;

71

mutations in the coding regions of the *lin-48* or *shavenbaby* genes would alter the expression of many downstream genes throughout the expression domains of the *lin-48* and *shaven-baby* genes. This asymmetry means that mutations in *cis*-regulatory regions will tend to have specific phenotypic effects, whilst mutations in protein coding regions will tend to have pleiotropic effects.

The evolved differences in excretory duct position and trichome pattern resemble the kinds of small phenotypic changes whose accumulation, evolutionary biologists have imagined, result in larger-scale differences between species and higher taxa. In both cases, multiple mutations with specific, non-pleiotropic effects caused evolutionary changes. Given the surprising complexity underlying these seemingly minor evolutionary differences, we must anticipate that a full accounting of the mutations that generate the multitude of differences between any two species, let alone between higher taxa, will tally a vast number of mutations.

The previous discussion implies that the pleiotropic roles of genes can bias which mutations contribute to phenotypic evolution. The pleiotropic roles of genes can also influence the evolutionary fate of duplicated genes. Gene duplication is the wellspring of most new genes, but most duplicated genes go extinct shortly after they first appear in a population for one of two reasons. First, the duplicate gene may be deleterious, in which case purifying selection will act against the functional duplicated allele. Second, the duplicate may be neutral. For example, the duplicate may be redundant. In this case, a mutation that inactivates one copy of the gene may be fixed by drift,

72

leaving a ghost of its former self behind, a pseudogene. The vast majority of duplicated genes meet one of these two fates.

A duplicated gene may be retained in the population, of course, if the duplicate itself provides a selective advantage— say, for example, by boosting the level of gene transcription in each cell. This almost certainly is how large tandem arrays of some genes, such as those encoding ribosomal RNAs, are fixed in populations. Alternatively, a duplicate can be retained if one copy acquires a new adaptive function before other mutations disable the gene. Since deleterious mutations are probably far more common than are mutations that can impart a new adaptive function on a gene, this model is unlikely to explain the abundance of duplicated genes in nature.

While natural selection is unlikely to favor most gene duplicates immediately, it turns out that natural selection is not required initially to cause retention of a functional duplicated gene in a population. A process starting with genetic drift can preserve both copies of a duplicated gene, as long as the original copy of the gene had pleiotropic roles. This process is named the duplication–degeneration–complementation model, after the three simple steps required for it to work, and it is illustrated in Figure 4.16. I illustrate the pleiotropic roles of an imaginary gene in Figure 4.16 as multiple *cis*-regulatory regions, but, in principle, pleiotropic protein functions will also allow the model to work. First, a gene with a complex *cis*-regulatory region *duplicates*. Second, one or more of the enhancers are lost from each copy of the gene through *degenerative* mutations. Finally, the *cis*-regulatory information from one gene *complements* the function of the lost *cis*-regulatory information at the duplicate. As a result, each copy of the gene takes on a more specialized role than the original gene had, and development now requires both gene copies.

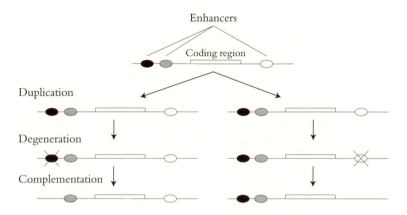

FIGURE 4.16. This figure illustrates the duplication–degeneration–complementation model of gene duplication. An ancestral gene, shown at the top, possesses multiple *cis*-regulatory enhancers. After duplication of this entire region, each duplicate accumulates mutations that eliminate the function of some of the enhancers. These mutations drift to fixation, since loss of function of each enhancer is complemented by the presence of a similar or identical enhancer in the duplicated gene. In a relatively small population, these neutral mutations can fix rapidly. In a large population, fixation of neutral alleles takes so long that additional deleterious mutations that eliminate gene function may accumulate on the same chromosomes containing the lost enhancers. These mutations will lead to complete loss of the gene duplicate, rather than maintenance of specialized duplicates.

One curious feature of this model is that it works best in small populations. In small populations, the degenerative mutations will behave more like neutral alleles, and neutral mutations fix more quickly in small populations than they would in large populations. The fixation of a neutral allele takes an average of $4N$ generations, where $N$ is the population size. In very large populations, neutral alleles thus take so long to fix that a chromosome with a mutation eliminating one enhancer is likely, by the time it fixes, to have accumulated a second mutation that has eliminated gene function. At this point, then, the

FIGURE 4.17. Comparison of the homeodomain sequences of the Hox3, Zerknüllt, and Bicoid proteins from flies reveals rapid evolution of the ancestral *bicoid* gene. Positions perfectly conserved between species for the Bicoid or Zerknüllt proteins are shown in black. Positions shown in grey boxes highlight conserved amino acids of the Bicoid or Zerknüllt proteins that have diverged since the gene duplication event. Of these amino acids, those amino acids of Zerknüllt that are conserved also with the Hox3 proteins are highlighted with grey boxes in the Hox3 sequences.

duplicated gene is dead and the mutation causing degeneration of the *cis*-regulatory module is irrelevant. The model therefore works best when degenerated neutral alleles drift through populations quickly. The two copies of the duplicated gene can then accumulate mutations that alter function in novel ways.

The evolutionary origin of the *bicoid* gene is consistent with the duplication-degeneration-complementation model. Two genes found in relatively recently evolved flies, called *zerknüllt* and *bicoid*, encode similar transcription factors that share sequence similarity with the *Hox3* gene of more ancestral fly species, as shown in Figure 4.17. It is likely that *zerknüllt* and *bicoid* represent duplications of the ancestral *Hox3* gene. In basal flies the Hox3 protein is expressed in maternal nurse cells and deposited in an egg. Later in development, Hox3 protein is expressed in the extraembryonic tissues of the developing embryo. In the more derived flies, the *zerknüllt* and *bicoid* genes appear to have divvied up the expression pattern of the ancestral *Hox3* gene, as shown in Figure 4.18. Zerknüllt protein is

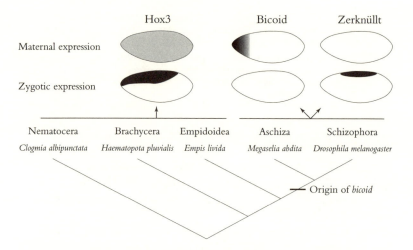

FIGURE 4.18. The genes *zerknült* and *bicoid* evolved as duplicated *Hox3* genes in the lineage leading to the Aschiza plus Schizophora. Ancestral dipterans contain a single *Hox3* gene. This ancestral gene shows maternal expression in the nurse cells and is deposited, apparently uniformly, throughout the oocyte. This gene is also expressed zygotically in the extraembryonic tissues. In lower diptera, these extraembryonic tissues extend to the anterior pole of the egg. In the more derived flies, the *zerknüllt* and *bicoid* genes have apparently divided the maternal and zygotic roles of the ancestral *Hox3* gene. Bicoid mRNA is expressed in nurse cells and deposited at the anterior pole of the egg, and Zerknüllt protein is expressed zygotically in the extraembryonic tissues. In the derived flies, the extraembryonic tissues do not extend to the anterior pole of the egg.

expressed in extraembryonic tissues. Bicoid mRNA is expressed in the nurse cells that provision the egg, and it is then transported to the anterior pole of the egg. The combined expression pattern of the *bicoid* and *zerknüllt* genes is similar to the expression pattern of the *Hox3* gene found in more basal dipterans.

Shortly after the duplication event, the *bicoid* gene evolved multiple amino-acid changes that conferred new molecular functions upon the Bicoid protein while the Zerknüllt protein remained similar to the ancestral Hox3 protein, as shown in Figure 4.17. The changes in the Bicoid protein allow it now to regulate the transcription of new target genes. In addition, the Bicoid protein can regulate the translation of mRNAs by binding the 3' untranslated region of specific mRNAs and by blocking the translational machinery. Thus, duplication of the *Hox3* gene and division of the ancestral expression pattern into two specialized patterns appears to have allowed the *bicoid* gene to evolve novel functions.

One striking feature of *bicoid* gene evolution is that evolution of novel functions required changes to amino acids that are conserved in many related genes. It is highly unlikely that these amino acid positions could have evolved within the related genes without disrupting many aspects of development. These dramatic changes in protein function could probably have occurred only in a duplicated gene, while the original gene retained its function.

## SUMMARY

The pleiotropic *roles* of a gene should be distinguished from the pleiotropic *effects* of individual mutations. Pleiotropic roles of a gene can be identified when some mutations alter fewer phenotypic characteristics than does a null mutation. The pleiotropic roles of genes result from the multiple functions of individual proteins and from differential gene expression. Thus, individual mutations in these genes may have many, few, or no pleiotropic effects. Theory predicts that mutations contributing to phenotypic evolution will tend to show few pleiotropic effects because the pleiotropic effects are likely to be deleterious. Several studies provide sufficient resolution of the muta-

tions contributing to phenotypic evolution to assess this prediction, and they support this prediction. Mutations in *cis*-regu-
latory regions can have few pleiotropic effects and may therefore contribute often to phenotypic evolution. Pleiotropy also
influences the probability that duplicated genes will be retained
in populations. Mutations that eliminate complementary functions of duplicated genes can result in selective maintenance of
an otherwise redundant gene duplicate, providing the opportu-
nity for evolution of novel gene functions.

# FIVE

# EPISTASIS

Genes not only *act* but also *interact*.

—Ernst Mayr, *Animal Species and Evolution*

Genes with complex interaction, though they may
be present in the population, are not selected
efficiently and hence remain at low frequency.

—James Crow, "Genetics of DDT resistance in *Drosophila*"

I f each distinct mutation had a predictable phenotypic effect,
no matter whether it occurred in one individual or another,
then dominance and pleiotropy would fully describe the range
of allelic effects. But separate mutations do not necessarily have
independent effects; they can have epistatic effects.

In traditional genetic terms, epistasis is said to occur when the
effects of one allele hide the effects of an allele at another locus.
For example, as shown in Figure 5.1, a null mutation of the
*Agouti* gene causes mice to have black fur. A null mutation of the
*Melanocortin 1 Receptor* gene causes mice to have yellow fur.
However, a mouse carrying null mutations of both the *Agouti*
and the *Melanocortin 1 Receptor* genes has yellow fur. The effect
of the *Agouti* mutation alone, black fur, is hidden by the effect of
the *Melanocortin 1 Receptor* mutation. One simple model to

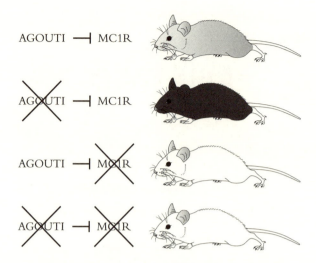

FIGURE 5.1. This figure presents a simplified illustration of epistasis between the *Agouti* and *Melanocortin 1 receptor* (*MC1R*) genes. A null mutation of *Agouti*, indicated with a big X, causes black fur. A null mutation in *MC1R* causes yellow fur. Mutation of both genes simultaneously causes yellow fur. The phenotypic effects of the doubly mutant mouse are best explained by a model in which the Agouti protein normally represses activity of the MC1R protein. Due to epistasis, the phenotypic effect of the double mutant is not the sum of the individual effects of each mutation alone.

explain this epistasis is that the two genes act in the same pathway, with the Agouti protein normally repressing the activity of the Melanocortin 1 Receptor protein and the Melanocortin 1 Receptor protein normally promoting the production of black pigment. The combined activity of both genes normally causes brown fur.

Epistasis tests provide a powerful method for detecting interactions between gene products during development. In the case of the *Melanocortin 1 Receptor* and *Agouti* genes, epistasis of the null mutations results from the fact that the protein products of these genes normally make direct contact. The Melanocortin 1

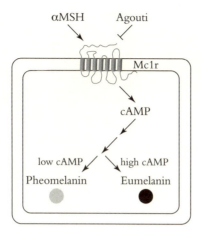

FIGURE 5.2. Pigment production in melanocyte cells is regulated through a pathway mediated by the Melanocortin 1 Receptor (Mc1r) protein. The Melanocortin 1 Receptor protein resides in the cell membrane and receives activating signals from α-Melanocyte Stimulating Hormone (αMSH) and repressing signals from the Agouti protein. Activated Melanocortin 1 Receptor protein causes increased cyclic-AMP (cAMP) levels within the cell, which ultimately leads to increased production of eumelanin, or black pigment. Reduced activity of Melanocortin 1 Receptor protein results in reduced levels of cyclic-AMP and shifts pigment production toward pheomelanin, a yellow pigment (shown here as gray).

Receptor protein resides in the cell membrane of the pigment-producing cells called melanocytes, as shown in Figure 5.2. Signaling by the Melanocortin 1 Receptor protein increases production of black eumelanin. Agouti protein is secreted from other cells and binds to, and represses signaling from, the Melanocortin 1 Receptor protein. Loss of function of the *Agouti* gene leads to excessive signaling from the Melanocortin 1 Receptor protein, generating darkly pigmented fur. Loss of function of the *Melanocortin 1 receptor* gene causes a dramatic decrease in the signaling that normally generates eumelanin. Instead, the pigmentation pathway diverges to produce the light

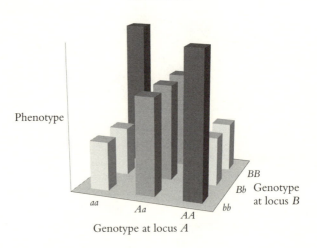

FIGURE 5.3. This figure illustrates a hypothetical example of strong epistasis, as measured in population genetics. The phenotype of an individual with a particular genotype for locus *A* depends on the genotype at locus *B*. Bars with the same shading represent genotypes that produce the same phenotype.

pigment pheomelanin. With loss of function of the *Melanocortin 1 Receptor* gene, the presence or absence of Agouti protein is irrelevant. The *Melanocortin 1 Receptor* gene acts "downstream" of the *Agouti* gene and loss of function of the *Melanocortin 1 Receptor* gene hides the effect of mutations in genes that act "upstream."

Epistasis tests using null alleles lead to relatively straightforward interpretations of genetic interactions. In nature, however, few null alleles segregate at high frequency in populations. Nonetheless, alternative alleles at some loci often modify the phenotypic effects of alleles at other loci. In population genetics, modification of allelic effects by variation at other loci is also called epistasis, and it can be detected when the effect of combining alleles at different loci in a single individual does not equal the sum of the effects of each allele alone. In the hypothetical example shown in Figure 5.3, the effect of one allele at one locus is dependent on the genotype at a second locus. By comparing the phenotypes in

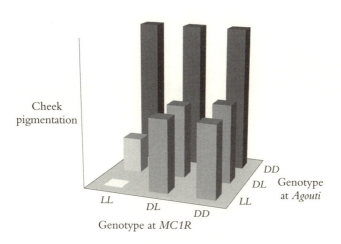

FIGURE 5.4. Naturally occurring alleles of the *Melanocortin 1 Receptor* (*MC1R*) and *Agouti* genes display epistasis. Alternative alleles are represented as *D* for Dark and *L* for Light.

the first and last row of Figure 5.3, we see that the effects of the *aa* and *AA* genotypes are reversed, depending on whether the individual also carries the *bb* or the *BB* genotype at locus *B*.

Epistasis in the population genetics sense has been detected in a wide variety of organisms in nature. For example, a population of oldfield mice (*Peromyscus polionotus*) from Florida contains alternative alleles at both the *Melanocortin 1 Receptor* gene and the *Agouti* gene. These naturally occurring alleles display epistasis for some pigmented areas on the mouse body. Pigmentation of the cheeks shows a striking pattern of epistasis, as shown in Figure 5.4. Animals homozygous for the *Agouti* allele conferring dark fur all have dark cheeks, independent of the genotype at the *Melanocortin 1 Receptor* gene. In contrast, individuals carrying at least one *Agouti* allele conferring light fur produce either intermediate or light cheek fur, depending on the genotype at the *Melanocortin 1 Receptor* gene.

∽

There are two dramatically different ways to think about the role of epistasis in evolution. First, multiple loci may simultaneously possess alternative alleles in a population, as in the population of oldfield mice discussed above. Second, mutations may substitute in a population and display epistasis with respect to mutations that substituted in the past. This serial epistasis can be detected only if the alternative alleles at each locus are tested in various combinations—for example, in an experimental cross.

Serial epistasis implies that history matters. The phenotypic effect of a new mutation is dependent on mutations that were substituted in the past. Consider again the multidimensional model of adaptation discussed in Chapter 4. Early in an adaptive walk, a mutation of large effect may be adaptive, but, late in the walk, the same mutation of large effect would not.

We also can consider the effects of serial epistasis over much longer time scales. Imagine a hypothetical protein-coding mutation that alters beak length in a warbler. The same mutation in the homologous gene of a snake might alter craniofacial development, but obviously not beak length. To take an even more absurd comparison, consider the human homologue of the fly *Ultrabithorax* gene. Mutations in *Ultrabithorax* can alter the haltere size in flies, but mutations in the human ortholog will certainly not influence the size of any appendages associated with flight.

Many authors prefer to use the term "developmental constraints" to describe these historical dependencies, but, at a population-genetic level, developmental constraint appears to reflect simply a long history of serial epistasis. The spectrum of phenotypic variation that may be generated by new mutations is dependent upon the genome that harbors the mutations. This is an obvious observation, but it has profound implications. Diversity begets diversity not, as Darwin suggested, in infinite variety, but along paths that were, to some extent, determined by the evolutionary paths already taken.

While it is easy to *imagine* how serial epistasis constrains evolution, it is much more difficult to prove that it did. Looking back across the vast stretches of evolutionary time, it may be difficult to determine whether alleles at two loci were fixed in series or whether alleles at these loci segregated in a population contemporaneously. By performing experimental evolution, we can sometimes observe the effects of serial epistasis. In these experiments, a sample of individuals is exposed to a novel environment and the population evolves improved fitness in this new regime. When these experiments are performed with microorganisms, such as bacteria, viruses, or yeast, it often is possible to follow changes in fitness over time. New adaptive mutations tend to sweep through populations relatively quickly, so fitness improvements usually result from the serial fixation of new mutations. Sometimes, replicate populations evolve to different maximal fitness levels, implying that different populations have followed different mutational trajectories to improved fitness and that some paths lead to higher fitness than others. This suggests that serial epistasis can influence the evolutionary trajectory of microorganisms. Presumably, serial epistasis is even more important for the evolutionary trajectories of more complex organisms, in which larger genomes encode a greater number of complex genetic interactions.

Serial epistasis plays another role in evolution. Consider again the multidimensional model of adaptation. A mutation may increase fitness by improving the phenotype with respect to one dimension, but the mutation may simultaneously degenerate the phenotype along other dimensions. New mutations may then arise that ameliorate these deleterious effects. For example, between 1958 and 1980, Australian sheep farmers applied large quantities of the insecticide diazinon in an attempt to control populations of the Australian sheep blowfly. Blowflies resistant to diazinon were first detected in 1965, and the resistance allele spread quickly through most of the population. While this mutation conferred resistance, it also reduced fitness in comparison with the original allele when flies were reared in the absence of insecticide. In the late 1960s, a second mutation at a separate gene that ameliorated the deleterious fitness effects of the first mutation spread through the population. Thus, serial epistasis can mitigate pleiotropic deleterious fitness consequences of otherwise adaptive mutations.

Epistasis implies that two mutations somehow alter the molecular activity of genes so that the effect of both mutations together is not simply the sum of the effects of each mutation alone. How exactly do molecular changes cause epistasis? Here is one example of serial epistasis for mutations that were substituted during the evolution of the vertebrate *Glucocorticoid Receptor* gene. Approximately 470 million years ago, a gene duplication in the vertebrate lineage generated two genes that evolved into the *Glucocorticoid Receptor* gene and the *Mineralocorticoid Receptor* gene, as shown in Figure 5.5. These genes encode receptors that act as transcriptional regulators when bound by corticoid steroid hormones. The Glucocorticoid Receptor protein binds the adrenal steroid cortisol and is required in humans for glucose homeostasis and for proper regulation of the stress response, inflammation, and immunity, as well as for other

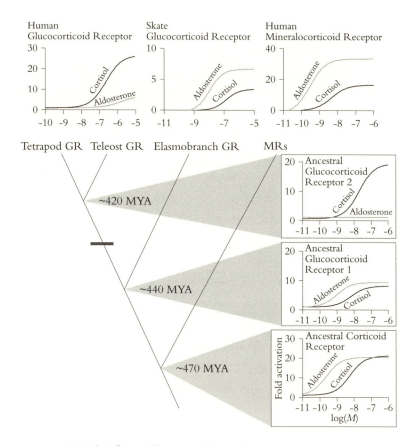

FIGURE 5.5. This figure illustrates the evolutionary relationship between two genes, the Glucocorticoid Receptor gene and the Mineralocorticoid Receptor gene. These two genes were derived from a gene duplication event that occurred approximately 470 million years ago. The vertebrate Glucocorticoid Receptor protein has high specificity for cortisol (*dark line*), while the skate Glucocorticoid Receptor protein, the human Mineralocorticoid Receptor protein and the resurrected ancestral Glucocorticoid Receptor 1 protein are more strongly activated by aldosterone (*light line*). This observation suggests that the ancestral protein had higher specificity for aldosterone. The graphs show the activation of receptors by different concentrations of ligands (*M*). The phylogenetic branch along which specificity of the Glucocorticoid Receptor protein shifted from aldosterone to cortisol is marked with a dark horizontal bar.

functions. The Mineralocorticoid Receptor protein, in contrast, is strongly activated by aldosterone, and it regulates electrolyte homeostasis, kidney and colon function, as well as other functions. These are the steroid specificities for the modern human Glucocorticoid Receptor protein and for the Mineralocorticoid Receptor protein. The different steroid specificities evolved sometime after the gene duplication event. How?

Clues about the ancestral state of the Glucocorticoid Receptor protein come from studies of skates, which do not produce aldosterone. As shown at the top of Figure 5.5, the skate Glucocorticoid Receptor protein is more strongly activated by aldosterone than by cortisol. This pattern is similar to the activation pattern of the human Mineralocorticoid Receptor protein and unlike the activation pattern of the human Glucocorticoid Receptor protein. This implies that the ancestral receptor, preduplication, had greater specificity for aldosterone than for cortisol. The Glucocorticoid Receptor protein evolved greater specificity for cortisol during, or just prior to, evolution of bony fishes, between approximately 420 and 440 million years ago.

We can gain more insight into these historical events by resurrecting putative ancestral proteins, first, by estimating the ancestral state of each amino acid and, then, by synthesizing the protein in the laboratory. Studies of the activation patterns of these resurrected ancestral proteins, shown along the right of Figure 5.5, allowed identification of the individual amino-acid substitutions that contributed to the specificity of the Mineralocorticoid Receptor protein for aldosterone and of the Glucocorticoid Receptor protein for cortisol. The simplest part of this complex story involves two substitutions: the substitution of a proline for a serine at residue 106 (S106P); and the substitution of a glutamine for a leucine at residue 111 (L111Q). The resurrected Glucocorticoid Receptor of an early chordate, called Ancestral Glucocorticoid Receptor 1 in Figure 5.5, is more sensitive to aldosterone than it is to cortisol. Introduction of both substitution S106P and substitution L111Q switches this sensitivity, so that the resurrected Glucocorticoid Receptor protein of the ancestor of vertebrates, called Ancestral Gluco-corticoid Receptor 2 in Figure 5.5, is more sensitive to cortisol than it is to aldosterone. But neither substitution, on it own, causes the sensitivity of the resurrected Glucocorticoid Receptor to shift toward greater specificity for cortisol. How, then, did this switch occur?

The three-dimensional structure of the ancestral and derived Glucocorticoid Receptor proteins clarifies the likely evolutionary path. The L111Q substitution causes only a small change in the structure of the protein and causes little change in the sensitivity of the protein for aldosterone and cortisol. In contrast, the S106P substitution partially unwinds a helix and destabilizes the protein region that binds ligands, as shown in Figure 5.6. The S106P substitution, on its own, therefore, impairs binding by both aldosterone and cortisol. In the presence of the L111Q substitution, however, the unwinding caused by S106P allows the glutamine now present at residue 111 to make a hydrogen bond with the C17-hydroxyl of cortisol, stabilizing the receptor–ligand complex. Aldosterone lacks this hydroxyl group and therefore does not make this hydrogen bond and does not bind tightly to the Glucocorticoid Receptor. This is why both mutations together cause the Glucocorticoid Receptor to bind more efficiently to cortisol than to aldosterone. Thus, it is likely that the L111Q substitution occurred first, causing only a small change in hormone sensi-

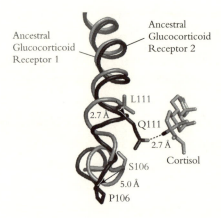

FIGURE 5.6. Detail of the three-dimensional structure of the resur-rected ancestral Glucocorticoid Receptor protein with (receptor 2) and without (receptor 1) the L111Q and S106P substitutions binding cortisol. The proline substitution at position 106 partially unwinds the helix and alters the position of the residue at 111. When glutamine is present at position 111, however, it can generate a hydrogen bond specifically with cortisol (and not with aldosterone). This stabilizes the receptor–ligand complex. Thus, the proline substitution at 106 can generate cortisol specificity only if glutamine has already replaced lysine at position 111.

tivity. The S106P substitution probably occurred later and flipped the specificity of the Glucocorticoid Receptor from aldosterone to cortisol.

In this case, serial epistasis did not result from direct interac-tion of the two amino acids that were substituted. Instead, epis-tasis resulted from a conformational change imposed by one amino acid substitution, which allowed an amino acid that sub-stituted earlier to make a particular hydrogen bond. We can presume that a large variety of molecular mechanisms can mediate epistasis.

～

Any given population, at any particular time, segregates for a small subset of all possible mutations. When a new mutation enters a population, the mutation may generate novel epistatic interactions. In some cases, the new mutation may generate epistatic interactions with other variants that previously had no phenotypic effect. That is, the new mutation may expose formerly "hidden" genetic variation. Hidden genetic variation can be detected experimentally by introducing a new mutation into a population. This experiment normally is performed by collecting organisms from the wild and inbreeding them for many generations to generate strains that are genetically identical at almost all loci. Each such strain represents a snapshot of the genetic variation from a population. A known mutation is then introduced into each strain, usually by introgression. For example, as shown in Figure 5.7, the extent of antenna to leg transformation caused by a mutation in the *Antennapedia* gene varies from strain to strain. The *Antennapedia* mutation reveals hidden genetic variation for antennal development.

The *Antennapedia* mutation causes a dramatic change in development, one that selection would likely never favor. But populations contain abundant hidden genetic variation for more subtle phenotypic changes as well. For example, as shown in Figure 5.8, a single copy of a null allele of the *Ultrabithorax* gene increases haltere size slightly. But, as shown in Figure 5.9, the *Ultrabithorax* null allele induces widely variable increases in haltere size in different fly strains. This hidden genetic variation was revealed only in the presence of the *Ultrabithorax* null allele.

Consider the fate of a new adaptive mutation from the time it arises in a population to when it either goes extinct or

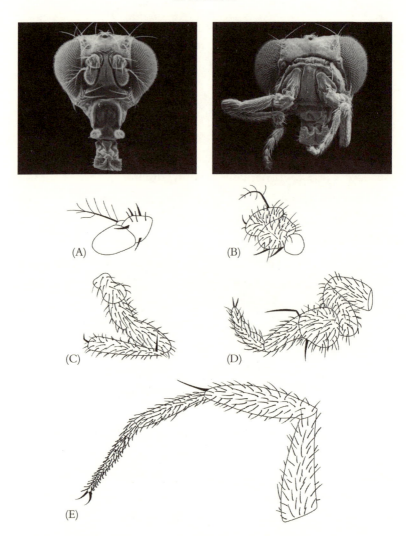

FIGURE 5.7. Hidden genetic variation is revealed upon introduction of a mutation in the *Antennapedia* gene into multiple wild isolates. Scanning electron micrographs of heads from a normal *Drosophila melanogaster* fly (*top left*) and a fly carrying a dominant mutation of the *Antennapedia* gene causing the development of legs in place of antennae (*top right*). (A) Drawing of a dissected antennae from a normal individual. (B–E) Drawings of antennae dissected from different wild isolates after introgression of the *Antennapedia* mutation.

Wild-type
(+/+)

Heterozygote
(*Ubx⁻*/+)

FIGURE 5.8. Images of a wild type haltere (*left*) and a haltere from a fly possessing one copy of an *Ultrabithorax* null allele (*right*). The fly carrying one *Ultrabithorax* null allele produces an enlarged haltere that often produces several bristles (*arrow*).

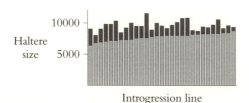

FIGURE 5.9. Haltere size, in arbitrary units, of females from introgression lines without an introduced mutation (*gray*) and possessing one copy of an *Ultrabithorax* null allele (*black*). Introgression lines are ordered from smallest to largest haltere size without the *Ultrabithorax* null allele. The *Ultrabithorax* null allele increases haltere size in all lines, but it increases haltere size to different extents in different introgression lines.

becomes fixed in the population. As the mutation traverses the population, it may encounter its own unique spectrum of epistatic interactions.

So far we have explored epistasis for observable phenotypic characteristics, such as coat color or pesticide resistance. Fitness, in contrast, integrates over the performance of the entire organism. Epistasis for fitness therefore can result from nonadditive fitness effects of any two phenotypic characteristics. Thus, genes *without* molecular or developmental interactions can exhibit fitness epistasis. In principle, any combination of phenotypic characteristics could interact to cause fitness epistasis. Here is an example from garter snakes.

Some garter snakes possess longitudinal stripes along their body and some display a heterogeneous spotty pattern. Individuals with intermediate patterns can be found in natural populations. Individuals also display variation for their escape behavior. Some escape by slithering in a straight line, while some perform many reversals. Intermediate escape behavior also can be found.

The fitness of animals performing a particular behavior depends on their pigmentation pattern. Striped snakes that escape in a straight line have higher survival rates than striped snakes that perform many reversals. Spotty snakes that perform many reversals have higher survival rates than spotty snakes that escape in a straight line. This may result from the fact that different patterns exhibit different optical properties when they are moving. For example, it is difficult to detect motion and to judge the speed of a striped pattern that is moving in a straight line. In any case, selection on behavior and pigmentation results in epistasis for fitness. As shown in Figure 5.10, this epistasis can be visualized in a three-dimensional plot of stripedness on one axis, behavioral reversals on another, and fitness on the third axis.

This clear example of epistasis for fitness involves two traits that almost certainly have no developmental connection. Many mutations for pigmentation patterns are unlikely to influence escape behavior. It seems more likely that independent genetic pathways control these two phenotypic traits. Nonetheless, the alleles underlying the variation for pigmentation and behavior

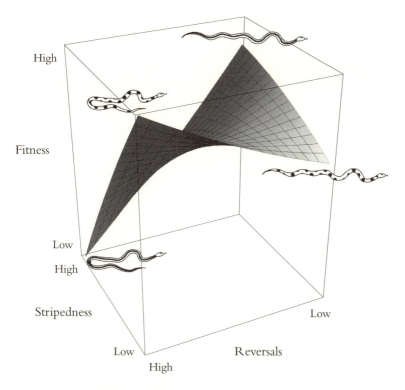

FIGURE 5.10. Epistasis for fitness in a garter snake. Natural populations vary in pigmentation patterns—striped versus spotty—and in behavior. Striped snakes survive at a higher frequency if they slither in a straight line. Spotted snakes survive best if they perform many reversals.

are trapped together in a kind of dance caused by epistasis for fitness. The presence of allelic variation for behavior prevents fixation of alleles for pigmentation, and vice versa. Genes with no developmental connections can therefore influence each other's evolutionary fates.

Fitness epistasis results not only from unusual circumstances, as in the balancing act of snake behavior and pigmentation, but it may occur any time natural selection favors an intermediate

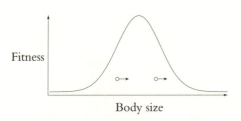

Fitness

Body size

FIGURE 5.11. A mutation that predictably increases body size shows epistasis for fitness when stabilizing selection acts on body size. When the mutation is present in an individual of below-average body size, indicated by the increase in size of the circle on the left, the allele increases fitness. When the mutation is present in an individual of above-average size, as for the circle and arrow on the right, it decreases fitness.

phenotype, such as an intermediate organ size or color. In most species, most of the time, for example, intermediate body sizes provide highest fitness and extreme body sizes perform worse. Selection for intermediate values is called stabilizing selection. As shown in Figure 5.11, any allele that increases body size will increase fitness only when the allele finds itself in a genetic background that tends to make the organism small. The allele decreases fitness in genetic backgrounds that tend to make the organism large. That is, stabilizing selection causes alleles that influence body size to show epistasis for fitness. Since most populations are reasonably well adapted to their current environment, most traits will experience stabilizing selection. Therefore, most alleles that contribute to most phenotypic traits in natural populations experience epistasis for fitness.

Does it matter that epistasis for fitness is abundant in natural populations? What are the potential evolutionary consequences of fitness epistasis resulting from standing variation?

Epistasis of standing variation is widespread in all natural populations, and it contributes to every imaginable phenotype. The examples discussed earlier represent only the smallest tip of the iceberg of epistasis in natural populations. The prevalence of epistasis suggests that it has the potential to play some role in evolution. The question is, what role or roles?

First, I must clarify that there is general agreement that serial epistasis has played a widespread and critical role in generating biological diversity. In contrast, there is little consensus about the potential role of epistasis of standing variation during evolution. One view, originally promoted by Ronald Fisher, is that epistasis of standing variation plays an insignificant role in adaptation, since natural selection cannot act efficiently on epistatic variation. The obvious potential for epistasis to generate novel variants, however, has attracted adherents repeatedly—most famously, Sewall Wright—to the proposal that epistasis of standing variation provides a key source of phenotypic novelty in natural populations.

I call attention to a different issue. In one respect, epistasis and pleiotropy may have similar evolutionary consequences. A mutation with pleiotropic effects may alter multiple phenotypic features during the same stage of life; or, the pleiotropic effects may manifest themselves at different life stages. A pleiotropic mutation may even have different phenotypic effects when the organism experiences different environments. In an analogous way, fitness epistasis reflects the different fitness effects of a mutation in different genetic backgrounds. As a mutation is passed from one generation to the next, each copy of this mutation experiences a different spectrum of alleles at other loci in the genome. Just as pleiotropic mutations have a lower probability of fixation than nonpleiotropic mutations, mutations displaying extensive epistasis with standing variation may have a lower probability of fixation. By increasing the variation in fitness associated with a particular mutation, epistasis effectively reduces the selection coefficient associated with a mutation. It is possible, then, that mutations that increase fitness—even mutations showing little or no pleiotropy—may have a lower probability of substituting in a population if they happen to display fitness epistasis with standing variation than if they do not.

It is not yet known whether mutations fixed by natural selection tend to show little epistasis with standing variation. There are many examples of presumptive serial epistasis between the mutations that fixed during evolution, but the existence of serial epistasis does not imply that those mutations experienced epistasis with standing variation. This is because standing variation may have different characteristics than adaptive mutations that have substituted during evolution, a point I discuss in more detail in Chapter 8. Because natural selection tends to favor mutations that confer the strongest fitness advantage, adaptive mutations that have substituted may show different patterns of dominance, pleiotropy, and epistasis than the standing variation.

## SUMMARY

In traditional genetics, epistasis—when a null allele at one locus hides the effect of a null allele at another locus—provides a phenotypic assay that can be used to detect molecular interactions between gene products. In population genetics, epistasis is measured as deviations from additivity caused by interactions between variable loci. Many alleles fixed in populations modify the effects of previously fixed alleles. This serial fixation of epistatic alleles appears to be a common mode of genetic evolution, and it helps to explain why the phenotype of a species can be dependent on the precise series of substitution events during the history of a species. In addition, natural populations contain many segregating alleles that display epistasis. Natural populations also contain abundant "hidden" genetic variation—phenotypic variation caused by epistatic interactions of standing variation with a newly introduced mutation. Epistasis for fitness can result from genes that interact at the molecular level and also, counterintuitively, from genes that do not interact, since fitness integrates over all phenotypic traits. Even stabilizing selection can cause fitness epistasis. Under some circumstances, fitness epistasis may thwart adaptive evolution, so adaptive mutations without epistatic effects may fix more often than will mutations with epistatic effects.

SIX

# POPULATIONS AND NATURAL SELECTION

Evolution in large random-mating populations,
which is recorded by palaeontology, is not
representative of evolution in general, and perhaps
gives a false impression of the events occurring in
less numerous species. It is a striking fact that none
of the extinct species, which, from the abundance
of their fossil remains, are well known to us, appear
to have been in our own ancestral line. Our
ancestors were mostly rather rare creatures. "Blessed
are the meek: for they shall inherit the earth."

—J. B. S. Haldane, *The Causes of Evolution*

The earlier chapters provided most of the conceptual framework required to integrate development and evolution, but we have not yet incorporated a realistic view of natural populations. Real populations exist in dynamic environments. Population sizes may increase or decline over time, and different species have different population sizes. For example, predators usually have smaller populations than their prey. Furthermore, as Darwin emphasized, there is a constant struggle for survival and reproduction. But, in this constant struggle, the targets of natural selection change frequently, and the strength of selection may change rapidly and often. An ice storm provides a brief, intense episode of selection for cold-resistance. A new pathogen supplies a novel source of selection. Variation in population size, differences in population structure, and fluctuations in the intensity and direction of selection provide the context for much of evolution. Here I review some of the consequences of these facts.

Population size is important, first, for the simple reason that mutations occur in individuals. Large populations, therefore, accumulate more new mutations in each generation than do small populations. Thus, large populations have a wider spectrum of mutations available for natural selection to act upon.

When we wish to evaluate the role of population size in evolution, it is not enough simply to consider the census population size—the observed number of individuals in a population. To see why, let's simulate genetic drift again by choosing marbles from a bag. We have ten marbles in the bag: five black, five white. We select one marble, note its color, and replace it. We repeat this exercise until we have sampled ten marbles. This mimics selecting gametes from a population to found the next generation. Sometimes we will select five white and five black marbles. Frequently we will select more than five white marbles or more than five black marbles. So, in this small population, for each generation, the frequency of black marbles can vary widely from, for example, 50% to 60% to 40%.

Now imagine a bag with 1,000 marbles: 500 white, 500 black. Resampling marbles from this bag will again cause the frequency of black marbles to fluctuate around 50%, but it is very unlikely to jump as high as 60% or as low as 40% in a single generation. Instead, it will tend to hover closer to 50%—moving, say, to 50.4% or 49.8%. Thus, genetic drift is more important in small populations than in large ones.

In this thought experiment, all marbles had an equal chance of being picked. In a real population, this would be equivalent to saying that each individual had an equal chance of contributing to the next generation. But, in real populations, for a slew of reasons, some individuals may far outreproduce others. Male sage grouse, for example, gather in large groups called leks to display to females in the hopes of securing one or more mates. The females then stroll about the display grounds to select a mate. Typically, only a few males garner the vast majority of matings. Males show a large variance in reproductive success: some mate with many females, and most mate with none. In this case, the number of individuals contributing to the next generation is much smaller than the census size of the population. From our marble experiment, we know that this effectively smaller population experiences more genetic drift than would a population in which all individuals reproduced.

The "effective" number of individuals reproducing is sometimes a more evolutionarily relevant measure of population size than is the census size. If we could measure the reproductive success of every individual in a population, then we could directly calculate what is called the effective population size, which often is abbreviated as $N_e$. Rarely is this feasible. We can *estimate* the effective population size, however, by examining its effect on genetic drift. The effective population size is the size of a hypothesized ideal population—one that behaves like marbles in a bag—that would experience the same amount of genetic drift as observed in our real population. Estimates of the effective population size of natural populations are typically smaller, and often much smaller, than the census size.

The reduced genetic drift in large populations relative to small populations allows selection to be more discriminating in large populations. If the population is too small, then the randomizing force of genetic drift may overwhelm the directional force of natural selection. If selection for a new mutation is very, very weak, then genetic drift may overwhelm the evolutionary dynamics of alleles even in a reasonably large population. There are, therefore, two possible reasons why a mutation may be effectively neutral and end up drifting through a population: either the selection coefficient is too small, or the population size is too small. Flip this thought around, and now imagine what kinds of mutations are fixed by selection in small and large populations. In small populations, only mutations conferring a large fitness improvement can efficiently be selected. In large populations, even mutations conferring a miniscule fitness advantage can be selected.

The population size thus influences substitution of adaptive mutations in two ways. First, population size mediates the rate at which new mutations are introduced into the population. The number of new mutations introduced into a population in each generation is directly proportional to the number of individuals in the population. Small populations may, therefore, contain little genetic variation. In particular, small populations may be deficient for a very particular subset of mutations, mutations with specific—nonpleiotropic and nonepistatic—effects on the phenotype. There are probably fewer possible mutations with specific effects than mutations that cause pleiotropic or epistatic effects. Thus, since selection will act on whatever variation is available, selection in small populations may preferentially utilize mutations with pleiotropic and epistatic effects.

Population size also influences which mutations are substituted, not only which mutations are available. The theoretical estimates of the probability of substitution of a new mutation, discussed in Chapter 3, were approximations that assumed that weak selection operated in a large population. As population size becomes very small, these estimates start to break down, because many alleles will be fixed by genetic drift with a probability of $1/2N$. In very small populations, the probability of substitution by genetic drift, $1/2N$, may even exceed the probability of substitution by selection expected in a large population, which, as we saw in Chapter 2, is approximately equal to twice the selective advantage enjoyed by the heterozygote (approximately $2hs$). How can population size influence which mutations are substituted? Imagine that two mutations occur in a population. Both have the same effect on one particular aspect of the phenotype—say, they both increase wing length by the same amount—which provides a modest increase in fitness. But one of the mutations also has the pleiotropic effect of increasing leg length, which is disadvantageous and reduces the net selection coefficient for this mutation to zero. In a large population, the first mutation—the one without pleiotropic effects—will be favored by selection and has a probability of substituting of about $2hs$, while the second mutation—the one that increases both wing length and leg length—will experience only genetic drift and has a probability of substituting of $1/2N$. Now, imagine that both mutations arose in a population sufficiently small that the mutation with specific effects is effectively neutral. The probability of fixation for both mutations becomes very similar, and both have approximately an equal chance of substituting.

Thus, for several reasons, small populations may fix a higher proportion of mutations with pleiotropic and epistatic effects. First, if mutations of large effect are more likely to have pleiotropic and epistatic effects, then smaller populations—which will preferentially substitute mutations of large effect—

may accumulate more adaptive mutations with pleiotropic and epistatic effects. Second, mutations with specific effects may be less abundant in small populations, again leading to substitution of an excess of mutations with pleiotropic and epistatic effects in small populations. Third, in small populations, mutations with specific, but small, effects may not substitute at substantially higher rates than mutations with pleiotropic or epistatic effects. All together, these effects may generate striking differences in the kinds of mutations that substitute in small and large populations.

Population size also influences the standing variation. First, a greater number of mutations occur in larger populations, so larger populations contain more effectively neutral alleles drifting through the population than do smaller populations. Second, as population size increases, selection becomes increasingly efficient. One important result of this increased efficiency of selection is that deleterious mutations under mutation–selection balance are kept at a low frequency.

Standing variation can influence future evolution in two ways. First, if the environment changes, then the selection coefficient associated with mutations may change: previously neutral or deleterious alleles may become selectively advantageous; previously neutral or advantageous alleles may become deleterious. Second, new mutations may interact with standing variation to generate epistasis, which may effectively reduce the selection coefficient associated with the new mutation. A potentially adaptive mutation spreading through a large population may experience a more diverse spectrum of epistatic interactions than would the same mutation in a small population. There are two reasons for this. First, larger populations harbor more variation than smaller populations and, second, an

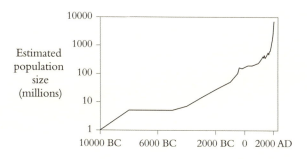

FIGURE 6.1. Estimated human population size over the past 12,000 years. The dip in the fourteenth century resulted from the Black Death pandemic. Note that population size is represented on a logarithmic scale.

adaptive mutation must pass through more individuals on the way to fixation in a large population than it must in a small population. On the other hand, deleterious mutations maintained under mutation–selection balance in small populations can have stronger fitness effects than can mutations maintained by mutation–selection balance in large populations. It is possible that these mutations of stronger effect generate higher levels of epistasis in small populations. It is thus not yet clear whether population size is expected to influence the spectrum of epistasis experienced by an adaptive mutation on its way to fixation.

Populations often change in size, and fluctuations in population size can influence evolution in several ways. So far in this book, I have assumed that population sizes were stable: the number of surviving offspring equals the number of adults. In this scenario, individuals replace themselves, genetically speaking, approximately once. Sometimes, however, populations grow rapidly. Spring brings a flush of vegetation, and insect popula-

107

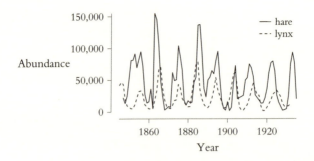

FIGURE 6.2. Lynx and their main prey item, hare, experience dramatic fluctuations in population size.

tions explode. A few migrants occupy a new island, and their descendants multiply to fill the land. As shown in Figure 6.1, over the past approximately 12,000 years, despite a few setbacks, including the Black Death in the fourteenth century, human population size has increased dramatically. From about 10,000 BC to 1700 AD, the human population size increased rapidly; and, since about 1700 AD, the human population size has increased even faster.

No species increases in population size indefinitely. Populations of many species, however, vary in size over time. Over the years, the population size of the lynx (*Lynx canadensis*) has tracked the population size of their favorite food, snowshoe hares (*Lepus americanus*), as shown in Figure 6.2. Also, climate changes can cause unpredictable changes in food availability. Major climatic fluctuations generated by El Niño events in the Pacific Ocean cause dramatic pulses of plant growth on the Galápagos Islands, leading to changes in the population sizes of Darwin's finches, as shown in Figure 6.3.

When the population size varies, the long-term effective population size is biased strongly downward by the generations of reduced population size. For example, the effective popula-

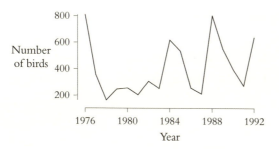

FIGURE 6.3. The number of individuals of the Darwin's Finch, *Geospiza fortis*, alive on the island of Daphne in January of each year fluctuated dramatically.

tion size—which counts only breeding birds—of the bird population illustrated in Figure 6.3 between 1976 and 1992 was estimated to be 197, which is considerably smaller than the total population size in all but one of these years. The reduced effective population size of populations that fluctuate in size is expected to reduce the efficiency of natural selection.

Individuals of a single species are rarely distributed homogeneously throughout a geographic region. Populations may be divided into multiple subpopulations in which individuals tend to mate with other individuals from the same subpopulation. When migrants only rarely make their way between the subpopulations, then the subpopulations will tend to behave like small populations. In some cases, subpopulations may thus experience local fixation of globally deleterious mutations by genetic drift. In addition, subpopulations may inhabit environments that impose different kinds of selection. Local selection may cause local fixation of alleles that would fare poorly when tested in other environments occupied by members of the species. If the species continues to occupy diverse habitats, then these locally adapted alleles are unlikely to spread through the entire population. Divergent selection in subpopulations of a species generates a conflict between the mutations favored at the local level and the mutations that may be favored throughout the species as a whole.

Mouse-ear cress, *Arabidopsis thaliana*, appears to have a subdivided population structure. The subdivision is enhanced by the fact that these plants are self-fertile, so that plants tend to mate not simply with their neighbors but mainly with themselves. Since many of these subpopulations inhabit different environments, the plants are exposed to different selection regimes. In general, populations that live in areas with mild winters experience selection for vernalization, whereas populations in more temperate areas and populations with harsh winters experience selection for rapid flowering. Selection for rapid flowering in subpopulations of *Arabidopsis thaliana* may have selected repeatedly for mutations in the *Frigida* gene. But most of these individual mutations are found in isolated populations and therefore remain at low frequency when considering the species as a whole. There is no evidence that any of these mutations are universally favorable for individuals of this species, and therefore no evidence that, in the future, one of these mutations—which all appear to be favored in local populations—will eventually substitute in the entire species.

Human subpopulations may have experienced local adaptation of alleles of the *Melanocortin 1 Receptor* gene that contribute to pigmentation variation between individuals. Four different mutations that alter the amino acid sequence of the Melanocortin 1 Receptor protein are associated with red hair or light skin color in humans. The mutations causing red hair disrupt the signaling capacity of the Melanocortin 1 Receptor

```
   Human ENALVVATIAKN...IFYALRYHSIVTLTRA...CNAIIDPLIYAFHSQ
    Bird ENLLVVAAILKN...IFYALRYHSIMTLQRA...CNSVVDPLIYAFRSQ
    Fish ENILVVAAIVKN...IFYALRYHNIVTLRRA...CNSVIDPIIYAFRSQ
Red or blonde      L           C        W           H
                  60          151      160         294
```

FIGURE 6.4. The mutations of the *Melanocortin 1 Receptor* gene that cause red or light hair in humans alter amino acids that are strongly conserved across vertebrates. Only part of the amino-acid sequence encoded by the *Melanocortin 1 Receptor* gene is shown. The bird sequence is from the chicken, *Gallus gallus*, and the fish sequence is from the pufferfish, *Takifugu rubripes*. Only the amino-acid positions altered in mutant human alleles causing red or light hair are shown, along with the position in the original human sequence.

protein, and at least one of these mutations may generate a null allele. As shown in Figure 6.4, these four mutations alter amino acids that are otherwise conserved across all other vertebrates. These patterns suggest that purifying selection has acted to maintain the same amino acids at these positions for hundreds of millions of years, but that human populations have recently accumulated mutations that alter these amino acids to generate phenotypic variation. Even though these mutations may confer a fitness advantage on their bearers in some geographic regions, it seems unlikely that these *Melanocortin 1 Receptor* mutations ultimately will substitute in the human species.

Looking back in evolutionary time—considering, say, the evolution of chimpanzees and humans from their common ancestor—there is a tendency to imagine that the differences between species represent the result of long-acting natural selection for whatever characteristics species currently display. We imagine, for example, that individuals in the lineage leading to modern humans experienced steady natural selection for increased brain size, upright posture, and less pronounced body hair. It is hard to say, even approximately, how accurate this model might be. Consider an alternative model. Natural selection for human features occurred in brief, strong episodes of selection, punctuated occasionally by selection in the opposite direction—for smaller brains, more hair. Perhaps natural selection acted in different ways on subpopulations, and most of this local adaptation is now lost. Perhaps the current average phenotype of humans represents the best compromise of serial epistasis that evolved in one of these subpopulations. The existence of multiple extinct hominine lineages suggests that we should not reject such a model out of hand.

In one respect, these details don't matter. Here we are now, chimpanzees and humans. Multiple related hominine lineages once existed contemporaneously with our ancestors and are now extinct. At the phenotypic level, this much is clear. But, at the genetic level, the details of the duration, strength, and dynamics of selection may matter a great deal. Once again, consider the geometric model for adaptation. Originally, we considered a population that had been displaced only a small distance from the optimum. Imagine now that the environment has changed dramatically and that the population has been displaced far from the optimum. If selection persists for a sufficient period, then we expect that the first mutations to substitute will have a relatively large effect on the phenotype. These may not be the best mutations for the job. They may, for example, have pleiotropic effects or experience epistatic interactions, but they may move the phenotype close enough to the optimum that they experience strong positive selection. If selection persists, then mutations of ever smaller effect will be favored. The strength of selection thus influences the phenotypic size of mutations that may be favored, and the duration of selection influences the distribution of mutational effects substituted.

Now what happens if the environment returns to its original state? Selection does not stop, it reverses. If some of the mutations selected during the first bout of selection have already substituted, then the population will evolve back toward the old optimum, either through the unlikely route of mutations that revert to the original DNA sequence or by novel mutations that move the mean phenotype of the population back toward where it was before. Of course, if the mutations that were positively selected in the first bout of natural selection had not yet substituted, then natural selection is likely simply to act against these alleles. Thus, the strength of selection, the duration of selection, and the dynamics of selection all influence which mutations ultimately substitute to generate differences between species.

While these are hypothetical examples, there is plenty of evidence that populations in nature experience fluctuating population sizes and variable selection pressures. Environmental conditions change over long and short time scales. Over the past 3 million years or so, approximately 100,000-year cycles of glaciation altered the available ecological niches dramatically and repeatedly on time scales similar to the time scales required to fix positively selected mutations. On a finer time scale, El Niño events create wild swings in the strength and direction of natural selection within the lifetimes of individual birds. As shown in Figure 6.5, populations of Darwin's finches experience repeated swings of strong selection, first for larger body size, and then for smaller body size, and then for larger body size. The same pattern has been observed for more specific aspects of the pheno-

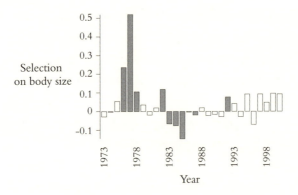

FIGURE 6.5. Selection on body size in the Darwin's Finch, *Geospiza fortis*, varies from year to year. Positive and negative values indicate selection for larger and smaller body size, respectively. Shaded bars are significantly different from zero.

type, such as beak size and shape. Presumably, mutations of relatively large effect on body size and on beak morphology are being buffeted to higher frequencies, then to lower frequencies, then back again on the time scale of decades.

The environment can change unpredictably, as described above for Darwin's finches, and it can change predictably. Dusk arrives at a predictable time every day, and the seasons change on a predictable schedule. All organisms have evolved to contend with predictable environmental changes such as these. For example, many organisms predict the coming of winter by detecting increasingly longer nights or the drop in average temperatures. These cues trigger changes in behavior, physiology, or development that improve survival over the winter, such as loss of leaves on deciduous trees or hibernation in rodents. Organisms also have evolved behavioral and physiological responses to less predictable changes, such as rainfall patterns. The ability of organisms to alter their phenotype to improve their chances of survival and reproduction is called phenotypic plasticity. Selection for phenotypic plasticity and selection in a constant environment may lead to selection of different kinds of mutations.

Many genes detect environmental changes and contribute to phenotypic plasticity. The Heat Shock Factor protein, for example, forms a trimer upon exposure to heat and this trimer then activates transcription of many genes that protect the cell from heat damage. At normal temperatures, the Heat Shock Factor protein remains a monomer and does not activate transcription. The Heat Shock Factor protein is, therefore, part of one mechanism that many organisms have at their disposal for dealing with unpredictable changes in temperature. Similarly, after exposure to cold days, transcription of the Flowering Locus C protein in *Arabidopsis thaliana* is downregulated, which allows the induction of flowering. The Flowering Locus C protein is part of the mechanism for optimizing reproductive output in a temporally variable environment. Phenotypic plasticity is so important—because environments are so tremendously variable—that the mechanisms regulating plasticity are likely to permeate every aspect of the organism. The mechanisms of plasticity involve not simply "plasticity" genes. Possibly all genes are involved in regulating plasticity. For example, while the Heat Shock Factor protein forms a trimer and binds to specific DNA sites after exposure to heat, many genes contain the specific DNA sequences that allow binding of the Heat Shock Factor protein trimer. These genes are as entangled in the mechanisms of plasticity as is the *Heat Shock Factor* gene. To live is to express phenotypic plasticity.

Maintaining phenotypic plasticity comes at a cost. If an organism were to discover an Eden with a constant environment, then selection to maintain the mechanisms of plasticity would fall away. Mutations that destroy the mechanisms of plasticity would be selectively neutral and would drift to fixation. If maintaining plasticity incurs an energetic cost, such as the production of Heat Shock Factor proteins in every cell, then removal of the environmental imperative would result in selection favoring the dismantling of plasticity.

Since many genes, or possibly all genes, regulate phenotypic plasticity, then loss or alteration of plasticity should be added to the list of possible pleiotropic effects resulting from new mutations. For example, mutations that cause an adaptive phenotype without compromising plasticity may be favored over mutations that destroy the ability of a gene to contribute to phenotypic plasticity.

## SUMMARY

Natural populations experience variation in population size and variation of the environment. The historical effective population size influences levels of standing variation: smaller populations contain less variation. The effective population size regulates the efficiency of natural selection: populations of small effective size experience more genetic drift than do populations of large effective size. Therefore, selection is less efficient in populations of small effective size. Temporal changes in population size reduce the effective population size and lead to lower levels of standing genetic variation. Spatial subdivision of populations can lead both to increased genetic drift in subpopulations and to selection for locally favored mutations. Temporal variation in the strength and direction of selection may lead to unusual patterns of genetic evolution, and environmental variability selects for adaptive plasticity. Mutations that increase the fitness of their bearers without affecting plasticity are likely to be favored over mutations that compromise plasticity.

# SEVEN

# PATHWORKS

Examining the commonplace from a strange
viewpoint is a valuable eye-opener whether
practiced by the poet who stands on his head
better to see the beauty of a sunset, or by the
scientist trying to visualize a crystal lattice from the
electron's point of view. The object in either case is
to divest a phenomenon of its obscuring
incrustation of familiarity.

—Arthur Winfree, "The scientist as poet"

Natural selection often acts in similar ways in different species. Both sharks and whales, for example, experienced natural selection for swimming ability and converged on hydrodynamic bodies. Sometimes phenotypic convergence results from parallel evolution, when homologous genes evolve in similar ways. For example, similar amino acid changes in Opsin proteins have evolved in different species to cause similar shifts in wavelength sensitivity. Parallelism occurs also in *cis*-regulatory regions. In humans, the *cis*-regulatory region of the *lactase* gene has evolved multiple times to cause lactose tolerance.

Parallelism presents a paradox. Since every phenotypic trait results from the activity of multiple genes, we might have expected that many different genes would evolve to generate similar phenotypic changes. But there are numerous examples in which the mutations contributing to evolutionary change apparently are not distributed randomly amongst all the genes that contribute to the construction of a particular phenotypic trait. For example, as discussed in the earliest pages of this book, much of the natural variation in flowering time of *Arabidopsis thaliana* populations results from at least twenty independent null mutations in the *Frigida* gene, even though at least 80 different genes can be mutated to alter flowering time. Some genes appear to be favored targets of evolutionary change—I will call these genes "hot spots". To see why hot spots exist, we must explore the details of gene interactions.

Pathways provide the simplest way to represent gene interactions. In a pathway, each gene product acts upon a substrate or on a second gene or gene product, as shown in Figure 7.1. The pathway concept suffices for thinking about the activity of a small number of genes, but it fails to capture the full complexity of the many gene interactions that govern development. In reality, multiple pathways converge on single genes and single gene products influence many other genes, thus forming networks.

Figure 7.2 illustrates part of the network controlling one of the most important developmental decisions a yeast cell will ever make, whether to cease growth and form spores that can eventually go on to mate—or not. When food is abundant, yeast cells grow and divide; but when environmental conditions deteriorate, yeast cells initiate the sexual phase of their life cycle. Since the first step of this process involves the development of spores, this process is called sporulation. There are two primary components to the network controlling sporulation: signals and genes. Glucose, for instance, acts as a signal to regulate transcription of the gene *IME1*. It also is useful to think of

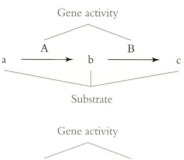

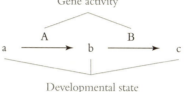

FIGURE 7.1. This figure shows several classical representations of pathways. At the top, I illustrate a classic biochemical pathway. Gene products act upon substrates to convert them from one state to a new state. In the middle, I have represented the transitions between developmental states in a similar way to a classic biochemical pathway. The activity of gene products converts cells from one developmental state to a new developmental state. The activity of gene product A alters the developmental state so that gene *b* is now expressed. The gene product B now alters the developmental state so that gene *c* is expressed. At the bottom, I illustrate a developmental pathway in a more simplified form as the action of one gene product upon another gene product or gene. For example, the product of the *a* gene may be a hormone that binds to the receptor encoded by the *b* gene which activates an intracellular signaling molecule encoded by the *c* gene. Alternatively, gene product A may induce expression of gene *b*. The product of the *b* gene then induces expression of gene *c*.

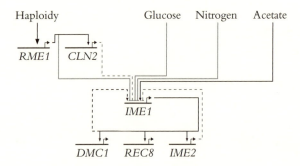

FIGURE 7.2. This illustration shows a simplified regulatory network controlling the developmental decision to sporulate in the yeast *Saccharomyces cerevisiae*. The horizontal line above a gene name contains the *cis*-regulatory region of the gene, not to scale. Solid lines emanating from genes represent proteins that act as transcription factors on other genes. Black lines indicate activation; gray lines indicate repression. Lines emanating from catch-all terms, like "Haploidy," represent signals produced by multiple intervening genes and gene products, which are not shown. Dashed lines indicate regulation through unknown mechanisms. The *CLN2* gene, for example, encodes a G1 cyclin, which presumably is not involved directly in transcriptional regulation by binding to the enhancer of the *IME1* gene.

gene products as signals. Thus, the IME1 protein serves as a signal that regulates transcription of other genes, such as *DMC1*, *REC8,* and *IME2*. As discussed in Chapter 3, these signals can activate or they can repress gene transcription. Transcriptional activation is illustrated with a line ending in an arrowhead. Transcriptional repression is illustrated as a line with a nail head at the end. To simplify, I sometimes summarize the activity of many genes and gene products by a line emanating from a catch-all description, like "Glucose" and "Nitrogen." In other words, in Figure 7.2, high levels of glucose in the growth medium repress transcription of *IME1* via other signals that are not explicitly illustrated.

It is easy to see that this network has a very different structure from the linear pathways shown in Figure 7.1. Perhaps the most striking thing about this diagram is that all of the information derived from environmental and internal signals is integrated at a single gene, *IME1*. The *IME1* gene appears to serve as a centralized decision point. It is the master regulator of sporulation. This centralized decision point can be seen only when we consider simultaneously the multiple inputs that are integrated during the decision to sporulate. If we were to examine only the role of one signal in regulating sporulation— for example, glucose—then we would reconstruct a linear pathway and perhaps overlook the central role of the *IME1* gene. Only by considering the network can we identify genes that occupy central decision points.

Some yeast strains sporulate quickly in response to environmental stress. These strains can be collected easily from oak tress and soil samples. In contrast, yeast strains isolated from vineyard fermentation vats, when exposed to similar environmental stresses, sporulate at low rates. Presumably, the abundant nutrients in vineyard vats have selected for high-growth strategies and low rates of sporulation. Most of the difference in sporulation efficiency between an oak-tree strain and a vineyard strain results from changes at the gene *RME1* and its transcriptional target, the *IME1* gene. A change in the *cis*-regulatory region of the *RME1* gene alters the amount of the RME1 protein that is produced, which alters regulation of the *IME1* gene. Two changes in the *IME1* gene alter sporulation rate, one in the protein-coding region and one in the *cis*-regulatory region. Other changes in other genes also contribute to sporulation rate. For example, a protein-coding change in the *RSF1* gene, which is not included in the network shown in Figure 7.2, also contributes to differences in sporulation rate. The RSF1 protein is a transcription factor that regulates mitochondrial genes. Respiratory signals also regulate transcription of the *IME1* gene, so it is likely that the mutation in the *RSF1* gene ultimately influences function of the *IME1* gene.

The structure of the sporulation network appears to affect which genes accumulate mutations that contribute to variation in sporulation rate. Mutations alter either the *IME1* gene itself or they alter genes that regulate *IME1*. No mutations altering sporulation have yet been found in any of the genes downstream of *IME1*. In addition, given the abundance of signals upstream of *IME1*, it is somewhat surprising that two important mutations are found in the *IME1* gene itself. Mutations influencing natural variation in sporulation efficiency appear to be overrepresented in the *IME1* gene and in the signals that regulate *IME1* activity.

Imagine how we might interpret the observation of these mutations without knowledge of the sporulation regulatory network, but with knowledge only of the fact that these genes somehow regulate sporulation. We might call all of the genes involved in regulating sporulation "sporulation candidate genes." We might feel satisfied that the mutations occurred in candidate genes for sporulation. But the regulatory network provides a more nuanced view of the data. We can see now that mutations do not appear to occur randomly in the regulatory network, but, instead, they appear to occur upstream of the master regulator *IME1* and in the master regulator itself. The available data suggest that by examining the network, we can start to *predict* where mutations will tend to accumulate during evolution.

The sporulation network illustrates an important and wide-spread principle of development. Regulatory information from many sources often converges on a single gene or gene product, encoding a developmental switch in the genome. Often, as for sporulation, much of the regulatory information is decoded at the level of the *cis*-regulatory region. The gene that integrates this information, and effectively makes the cell-fate decision, then regulates a set of downstream genes that carry out the required steps to perform the developmental event.

The location of a gene within a regulatory network is therefore likely to influence the pleiotropic effects of mutations in that gene. Imagine a single protein encoded by a gene—we'll call it *Activator*—that upregulates transcription of six other imaginary genes, *a, b, c, d, e,* and *f,* as in the following figure:

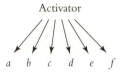

Any mutation that alters the way the Activator protein works may cause altered transcription of all six genes, *a* through *f*. In contrast, a mutation in gene *a* may change the way *a* is regulated by Activator protein, but this will not change the way genes *b* through *f* are regulated. Mutations that alter Activator protein expression or function will alter more regulatory interactions than mutations in only one target gene of the Activator protein.

While mutations that alter the function of the Activator protein may cause pleiotropic molecular changes, we do not yet have enough evidence to determine whether these mutations would cause pleiotropic phenotypic effects. Consider first the case where changes in gene *a* cause a specific phenotypic effect—say, a slightly wider wing—and changes in genes *b* through *f* cause other phenotypic effects, as in this figure:

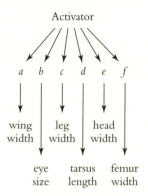

In this case, mutations that alter the function of the Activator protein will have more pleiotropic effects than will mutations in any of the downstream genes *a* through *f*. If selection favors change only in the width of the wing, and if any of the other phenotypic changes have any deleterious consequences, then selection will favor mutations in the *a* gene over mutations in the *Activator* gene.

What if changing expression of the *a* gene on its own has no effect? Instead, changing expression of genes *a*, *b*, *c*, *d*, *e*, and *f* results in a slightly wider wing, as in this figure:

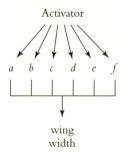

In this case, a mutation that alters the function of the Activator protein can cause a coordinated change in the expression of all six genes, *a* through *f*. Mutations in only one of these six genes may be insufficient to alter the phenotype. Changing the phenotype requires changes in the activity of all six genes. In this case, a mutation in the *Activator* gene is more likely to be favored by natural selection than are mutations in six independent genes.

Consider one more example, to round out the spectrum of possibilities. In this example, the Activator protein regulates only gene *a*, as in this figure:

Imagine that the Activator protein regulates *a* in the wing, in the leg, in the eye, and in the gut. Altering the way the Activator protein works, for example by changing the way the Activator protein binds to DNA, will alter the way the *a* gene is regulated in all of these tissues. Changes to the Activator protein will probably have pleiotropic phenotypic effects. Imagine, however, that either the *Activator* gene or the *a* gene can be transcriptionally repressed (by changes in an upstream gene product or by *cis*-regulatory changes) in one of these tissues. This change causes a specific phenotypic effect and would have a greater likelihood of being fixed by selection than would changes in the way the Activator protein regulates the *a* gene.

These three examples share a common thread: pleiotropic effects are determined by how regulatory networks shape the phenotype. In the first case, the phenotype that best matches the target of natural selection results from changes in only one downstream gene, *a*; changing the activity of the Activator protein would cause pleiotropic effects through other downstream genes. In the second case, the adaptive phenotype is caused only by coregulation of all six downstream genes; it is easier to change the regulation of just the *Activator* gene than the regulation of all six downstream genes to generate the desired phenotypic change. In the third case, changing the tissue-specific regulation of one gene generates a specific adaptive phenotype, whereas changing the way the gene products regulate each other generates pleiotropic effects.

Pleiotropy results from genetic networks. Thus, a complete map of gene regulatory networks might provide a useful guide to understand how genomes evolve. The regulatory network illustrated in Figure 7.2 captures many of the interactions controlling a single developmental decision of a single-celled organism. In a single cell, we can simply follow the events forward in time to generate an accurate picture of the regulatory decisions that lead to alternative cell fates. The situation becomes more complicated when we consider multicellular organisms. In a multicellular organism, different cells use different parts of the total regulatory network. At any one time in a single developing embryo, thousands of different cells may be pursuing different trajectories through a regulatory network. With so many different instantiations of the regulatory network pursued in different cells, there would seem to be little hope of identifying favorable locations in the network for evolutionary change. In the next section, I offer a way out of this tangle.

༄

For the single-celled yeast, we constructed a simple model of sporulation by working forward in time, from multiple signals to the genes that integrate these signals and drive sporulation. It likewise seems intuitive to model development of a multicellular organism as it happens, forward in time, from a small number of regulatory interactions in the fertilized egg to the deployment of batteries of genes to cause cell differentiation. We might attempt to include all regulatory interactions in all cells. In principle, this should allow us to reconstruct the logic that guides development.

For example, in Chapter 3, we explored how four transcription factors determine expression of the *even-skipped* gene in a single stripe in the embryo. However, these four interactions are embedded in a far more complex network. Figure 7.3 illustrates part of the regulatory network defining expression patterns of segmentation genes in the early *Drosophila melanogaster* embryo.

With practice, one can learn to examine such a diagram and extract a huge amount of information about how embryos develop. On the other hand, this view condenses the activity of many genes acting within thousands of cells over a span of time into one static image. This view obscures the individual regulatory interactions that generate each step of development.

We might solve the temporal visualization problem by illustrating the regulatory interactions active only over a particular span of time. For example, we could consider the regulatory interactions in only one row in Figure 7.3. But this still carries the disadvantage of summarizing the behavior of thousands of cells with potentially different fates in one image. Cells located in the anterior of the embryo express one set of genes, and cells

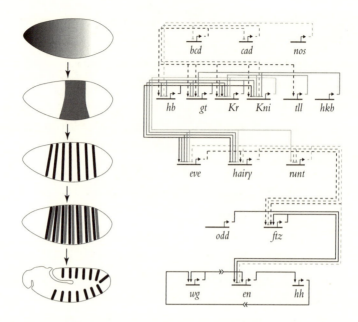

FIGURE 7.3. This figure illustrates a small part of the total regulatory network that establishes segments in the *Drosophila melanogaster* embryo. Along the left side, the spatial expression patterns of one or two genes (selected from those shown along the right side) are illustrated for several developmental stages. Along the right side, the regulatory interactions between some of the genes involved in development of the *Drosophila melanogaster* embryo are shown. A fraction of the regulatory interactions between genes are illustrated using symbols that were introduced in Figure 7.2. Most of these gene interactions are known to involve direct binding of transcription factors to the *cis*-regulatory region of each gene. The double arrows connecting *wingless* (*wg*) with *engrailed* (*en*) and *hedgehog* (*hh*) with *wingless* (*wg*) indicate intercellular signaling cascades. Other gene symbols are for *bicoid* (*bcd*), *caudal* (*cad*), *nanos* (*nos*), *hunchback* (*hb*), *giant* (*gt*), *Krüppel* (*Kr*), *Knirps* (*Kni*), *tailless* (*tll*), *huckebein* (*hkb*), *even-skipped* (*eve*), *odd-skipped* (*odd*), and *fushi-tarazu* (*ftz*).

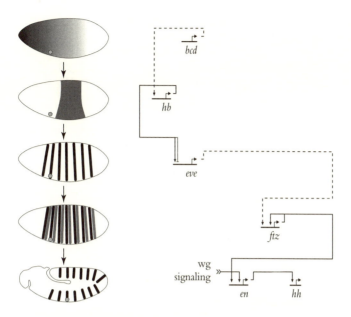

FIGURE 7.4. This figure illustrates the path through the network shown in Figure 7.3 that generates a single cell expressing Engrailed protein. The focal cell is illustrated as a small hexagon on the ventral side of each embryo.

in the posterior express a different, though possibly overlapping, set of genes. Each cell experiences only part of the network.

We could try to solve this problem by viewing development from the perspective of a single cell. For example, we could follow a single cell lineage through the network, as shown in Figure 7.4. This image illustrates that a single cell experiences only part of the entire regulatory network. Following this cell lineage forward in time partially solves the problem of how to visualize dynamic changes in space and time.

We are left with one final problem. As cells proliferate, we must choose, at each cell division, which path to follow. What motivates our choice? We do not necessarily know where we are headed, although following a random path will certainly take us somewhere. Usually, we wish to focus on a particular differentiated state; which developmental events generated the teeth, which ones generated the heart? Or, from an evolutionary perspective, we might ask what happened to cause one cell, or one group of cells, to evolve a new differentiated state: how did spots evolve on butterfly wings, how did birds evolve feathers? It may therefore be more useful to view development in reverse. Start with a final differentiated cell and trace the path backwards through the regulatory network to an earlier progenitor cell. This path defines a unique cell lineage. The path of a single cell lineage traced backwards in time captures only part of the full regulatory network. This simpler structure does not ignore, however, any of the regulatory interactions that created the differentiated cell. I call this a pathwork, because it captures the path through a network that generates a particular cellular phenotype. Whereas networks provide a genome-centric view of development, pathworks provide a cell-centric view. Pathworks work because cells are the units of development.

The concept of pathworks provides many advantages. Pathworks clarify that individual cells tend to make relatively simple cell-fate decisions at any particular time. Pathworks also clarify how particular mutations affect a single cell and how these mutations cause pleiotropic effects.

Imagine that we wanted to understand the regulatory interactions that cause a particular cell on the third abdominal segment of the *Drosophila melanogaster* first-instar larva to differentiate a trichome, say the particular trichome circled in Figure 7.5. Starting with the differentiated cell and moving backwards in time, the cell "first" deposited cuticle on the apical surface of the cell. This step required the regulation of multiple genes involved in making and depositing cuticle. In addition, tri-

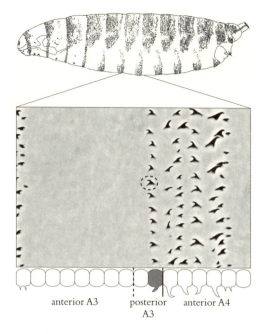

FIGURE 7.5. A single cell in the posterior compartment of the ventral third abdominal segment of *Drosophila melanogaster* is encircled with a dashed line. The drawing below illustrates the position of this cell, filled in gray, in the most posterior cell of the posterior A3 segment.

chomes are darker than surrounding cuticle, so this step involved regulation of pigmentation genes. These steps required activation of multiple genes:

In addition, trichome differentiation involved unique changes to the extracellular matrix, which required activity of additional genes:

Just prior to these steps, the cell adopted a pointed and recurved shape caused by the rearrangement of Actin protein within the cell. This step required expression of genes involved in Actin protein mobilization:

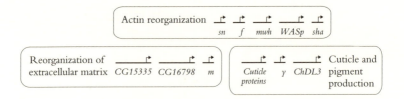

Current evidence indicates that at least some of these genes are directly regulated by the Shavenbaby protein. It is possible that all of these genes are directly regulated by Shavenbaby protein, but this requires further experimental validation. To keep it simple, I represent activation of all of these genes directly by Shavenbaby protein.

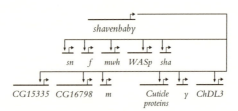

Transcription of the *shavenbaby* gene was upregulated by the activity of multiple upstream factors, including the transcription factor products of the *SoxNeuro* and *Dichaete* genes and potentially also downstream transcription factors of the Wingless and EGF Receptor signaling pathways. The Shavenbaby protein reciprocally activates transcription of the *SoxNeuro* and *Dichaete* genes to create a positive feedback loop:

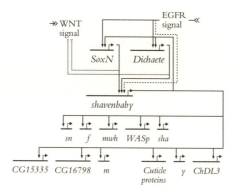

Transcription of the *SoxNeuro* and *Dichaete* genes may be activated by positive signals from the EGF Receptor signaling pathway. The SoxNeuro and Dichaete proteins also appear to repress signaling through the Wingless signaling pathway in this cell. The signal from the EGF Receptor pathway emanated from a cell just to the posterior of our focal cell, as shown in Figure 7.6. Production of this signal required expression of the Rhomboid protein in this more posterior cell. Transcription of the *rhomboid* gene was activated in this cell by positive signals from the Hedgehog signaling pathway. The Hedgehog signal was previously released from our focal cell. Transcription of the *hedgehog* gene in this cell was activated by the Engrailed protein. Thus, we can illustrate the pathwork for this cell from the initiation of *engrailed* gene expression to activation of multiple genes involved in terminal differentiation, as shown in Figure 7.6.

Moving backwards in time, the pathwork traces a path that led to this trichome-bearing cell. To fully appreciate the value of pathworks thinking, let's examine what happens after addition of one more cell to the diagram. For example, consider the cell positioned just to the anterior of the trichome-bearing

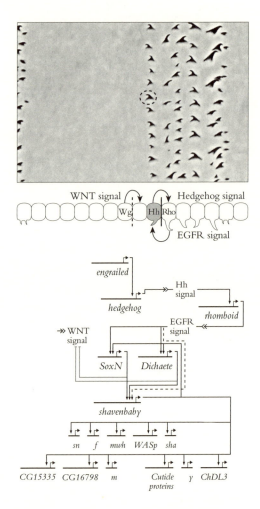

FIGURE 7.6. This diagram illustrates the pathwork for trichome production in the focal cell from the initiation of Engrailed protein expression. The focal cell, in gray, is illustrated at the top. The partial pathwork, from expression of Engrailed protein until activation of the developmental gene battery generating a trichome, is illustrated below.

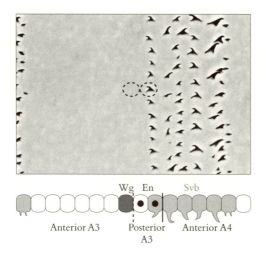

Wg En          Svb

Anterior A3        Posterior      Anterior A4
                      A3

FIGURE 7.7. Two cells in the posterior compartment, circled with dashed lines in the picture at top, differentiate into two different fates. Both cells express Engrailed protein (En: *black dots*). Of these two cells, only the posterior cell expresses Shavenbaby protein (Svb: *cells filled in light grey*). Wingless protein (Wg), expressed in the most posterior cell of the anterior compartment (*cell filled in dark gray*), signals to the anterior cell expressing Engrailed, which ultimately results in the repression of transcription from the *shavenbaby* gene.

cell, the naked circled cell in Figure 7.7. This cell does not differentiate a trichome, although it is positioned in precisely the same dorsal-ventral register as our focal trichome-bearing cell. Both the trichome-bearing cell and the naked cell develop within the posterior compartment, and both express the Engrailed protein. The naked cell does not, however, express the Shavenbaby protein. Transcription of the *shavenbaby* gene is repressed, potentially directly, by the Wingless signaling pathway and because transcription of both the *SoxNeuro* and *Dichaete* genes is repressed by the Wingless signaling pathway.

Addition of this cell adds some small amount of complexity to the pathwork, shown in Figure 7.8. The perspective from the

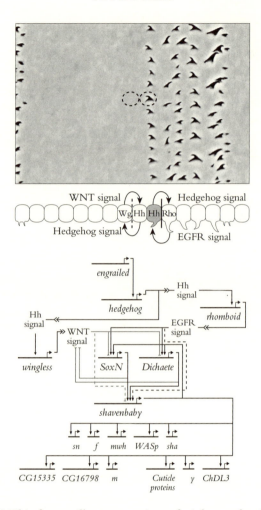

FIGURE 7.8. This figures illustrates a view of trichome development from the perspective of the genome. The simplified genetic network shown at the bottom contributes to determining the phenotypic characteristics of the adjacent cells that produce either naked cuticle or a trichome.

genome combines both cell fates. Without the spatial information, we can't discern precisely how this network generates two
different cell fates. By partitioning the pathworks of the two
cells into different cells, as in Figure 7.9, we see how the gene-
regulatory interactions in neighboring cells cause differentiation of naked cuticle versus a trichome. Each cell experiences
only a small part of the entire network. In addition, each cell
receives different quantitative levels of signals and expresses different levels of transcription factors. These analog variations can
be interpreted by single *cis*-regulatory regions to generate digi-
tal, on or off, outputs. The concept of pathworks provides a natural framework for envisioning the developmental effects of
quantitative variation in signals amongst cells.

Pathworks can illuminate why some genes are genetic
hotspots. Consider, again, the evolution of the *shavenbaby* gene.
In Chapter 4 we saw that multiple mutations in the *shavenbaby*
gene caused evolution of trichome patterns in *Drosophila sechellia*. The *shavenbaby* gene also has evolved in parallel in multiple
lineages to generate diversity in trichome patterns. Within the
*Drosophila virilis* clade of flies, which diverged from *Drosophila
sechellia* approximately 60 million years ago, trichome patterns
similar to the *Drosophila sechellia* pattern have evolved multiple
times, as shown in Figure 7.10. Current evidence suggests that
changes in the expression pattern of the *shavenbaby* gene caused
these changes in trichome patterns in the *Drosophila virilis*
clade. In a third lineage of flies, the olive fruit fly, *Bactrocera oleae*,
has evolved a pattern of trichomes similar to the pattern found
in *Drosophila sechellia*. This trichome pattern is precisely correlated with expression of the Shavenbaby gene product. Thus, in
three independent fly lineages, multiple evolutionary transi-

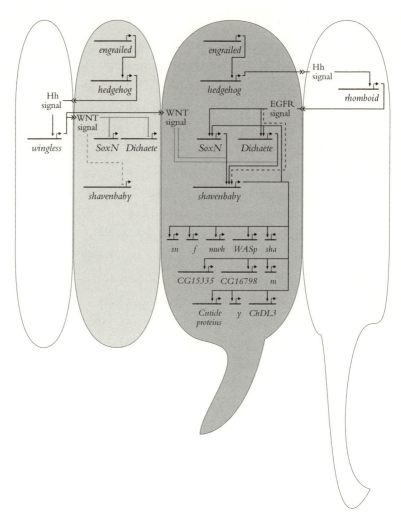

FIGURE 7.9. This figure illustrates a view of trichome development from the perspective of the cell. When pathworks are sequestered into their respective cells, a simpler pattern emerges than the one observed from the perspective of the genome, which was shown in Figure 7.8. Only the pathworks responsible for the fates of the central two cells, shaded in grey, are shown. Other pathworks determine the fates of the outer two cells.

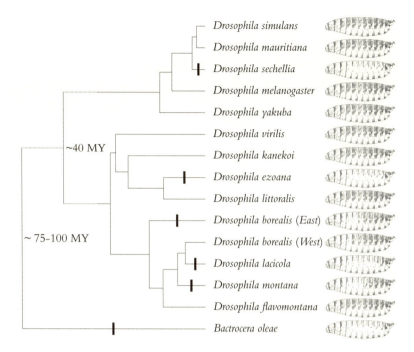

FIGURE 7.10. Parallel evolution of trichome patterns amongst fly larvae. A phylogeny of the species is shown on the left. Drawings of the trichome patterns of first-instar larvae are shown on the right. A mapping of trait evolution on the phylogeny using the principle of parsimony is illustrated and shows loss of trichomes in multiple independent lineages. An alternative equally parsimonious mapping requires loss of trichomes prior to the splitting of *Drosophila borealis (East)* and subsequent gain of trichomes in several lineages.

tions in trichome patterning are associated with changes in *shavenbaby* gene expression. The *shavenbaby* gene appears to be a hot spot for evolutionary changes in larval trichome pattern.

Two factors probably determine why the *shavenbaby* gene is a hot spot for trichome pattern evolution. First, the *shavenbaby* gene sits in a bottleneck position in trichome-differentiation pathworks. Figure 7.9 illustrates only some of the regulatory information integrated by the *cis*-regulatory region of the *shavenbaby* gene, but even this limited view illustrates that the *shavenbaby* gene integrates multiple signals. While we have focused on trichomes, these multiple upstream signals regulate other processes, such as neuron and muscle differentiation. For example, the *hedgehog* gene is required to establish the expression domains of both the *wingless* and *rhomboid* genes, as shown in Figure 7.8. Changes in the expression pattern of the *hedgehog* gene would have an impact upon other signaling pathways and would alter multiple aspects of the final phenotype. In contrast, changes in the expression pattern of the *shavenbaby* gene cause changes only in the pattern of trichomes. Thus, if selection acts to alter trichome patterns, then, amongst all genes in the trichome differentiation pathwork, changes in the expression of the *shavenbaby* gene will cause the fewest pleiotropic phenotypic effects.

There is a second potential reason why the *shavenbaby* gene is a hot spot. The Shavenbaby protein activates genes that drive an entire module of morphogenesis—the differentiation of a trichome. Changes in individual genes downstream of the Shavenbaby protein can cause slight changes in the shape of a trichome, but blocking the activity of any single downstream gene does not prevent differentiation of a trichome. Of course, simultaneously blocking the function of many of these genes could prevent trichome development, but changes in the *shavenbaby* gene alone can alter trichome patterns far more efficiently.

Thus, the *shavenbaby* gene sits at a convergence point in the trichome differentiation pathwork. Upstream components have known pleiotropic roles, even in a single cell. Downstream differentiation gene batteries are regulated en masse by the

Shavenbaby protein. The *shavenbaby* gene occupies a privileged position in this pathwork, where evolutionary changes minimize pleiotropy and maximize the phenotypic output.

Development, therefore, appears to bias genetic evolution toward genes that reside in integrative control points of pathworks. We saw this for sporulation in the single-celled yeast and for trichome patterning in the multicellular *Drosophila*. This is a strong claim, and one that awaits testing as additional pathworks are revealed and relevant phenotypic variation is dissected into its constituent mutations. This also is a specific claim regarding only the genetic means by which particular phenotypic traits evolve. The existence of hot spots *does not* necessarily imply that the phenotypic characteristics governed by these pathworks will evolve more easily than other phenotypic characteristics. But it may be worth giving serious consideration to this more controversial claim. It is possible, for example, that the ease of phenotypic evolution depends on the structure of pathworks underlying phenotypic characteristics. Pathworks that lack one or a few key control points may admit few mutations with limited pleiotropic effects, making them relatively resistant to evolutionary change.

Trichome development seems to provide the ideal scenario for identifying pathworks. Trichome-forming cells represent distinct populations of cells, and the key trichome-differentiation signal is a cell-autonomous developmental switch. It may be tempting, therefore, to dismiss the generality of pathworks. Perhaps pathworks break down when considering developmental decisions that are regulated by nonautonomous signals, such as hormones. This objection is partially overcome by recognizing that even the trichome pathwork for a single cell includes nonautonomous signals from neighboring cells, as shown in Figure 7.9. But the generality of pathwork thinking will be tested only by examining additional networks.

↶

In the previous section, I stressed that one key to identifying pathworks is to work backwards from the final differentiated state of a cell to earlier developmental decisions. The other key is to define precisely which differentiated state is of interest. For the previous example, we were interested in trichome differentiation and not in segmentation, even though these two processes share activity of many of the same genes. If we wanted to know whether the regulatory network influenced the evolution of segmentation, then we would have drawn a different pathwork. To define a pathwork, first define the differentiated state of interest or the precise phenotypic response of interest, as in the following example.

Flowering time in *Arabidopsis thaliana* is controlled by multiple independent signals, such as photoperiod, hormone signaling, and vernalization. Figure 7.11 illustrates much of the genetic network that transduces these signals to regulate flowering time. In its complexity, this diagram superficially resembles the network of gene regulatory interactions governing segmentation in *Drosophila melanogaster* that was shown in Figure 7.3. However, like the *Drosophila* segmentation network, the *Arabidopsis* flowering-time network encompasses many different kinds of processes. For example, flowering time is regulated by photoperiod through a part of the network that includes Phytochrome and Cryptochrome proteins. But flowering time is regulated also by a so-called autonomous pathway. Proteins active in the autonomous pathway promote flowering independently of photoperiodic cues. Flowering time also is regu-

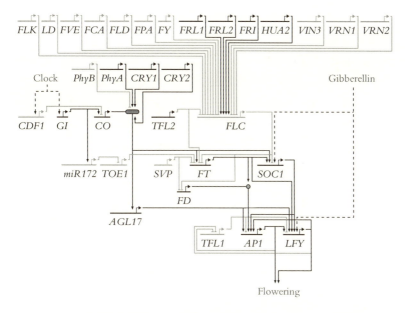

FIGURE 7.11. This figure illustrates much of the genetic network that regulates flowering time in the plant *Arabidopsis thaliana*. Multiple genes are involved in the photoperiodic clock and gibberellin portions of the network. Different genes in this network have, by tradition, been grouped into different "pathways." Along the top, from left to right, the genes *FLK* to *FY* are part of the Autonomous Promotion Pathway and the genes *FRL1* to *VRN2* are part of the Vernalization Pathway. The genes downstream of the clock and regulating the *FT* gene, including the *Phy* and *CRY* genes, are part of the Photoperiodic Promotion Pathway. The genes *TFL2*, *FLC*, *FT*, and *SOC1* have been called Floral Integrators. The genes *TFL1*, *AT1*, and *LFY* are called Meristem Identity genes.

lated by the amount of the hormone gibberellin circulating in the plant, and flowering time is regulated also by vernalization. Different parts of this network act at different times during the life of a plant and, to a certain extent, on different tissues. For example, long days induce flowering in mature plants. In contrast, vernalization acts on immature plants over the winter.

Figure 7.11 presents a good chunk of the flowering time regulatory network, but we have not yet determined which of these multiple processes are of interest. We therefore cannot yet draw a pathwork. We might ask, "Which genes have evolved to alter the vernalization response?" This question immediately allows us to draw the pathwork for vernalization, shown in Figure 7.12. This pathwork includes input from the autonomous pathway, since, as the name implies, the autonomous pathway is always active. This pathwork illustrates that the *Flowering Locus C* gene sits in an apparently important central control position for vernalization. Flowering Locus C protein is a transcription factor that acts to repress genes that normally promote flowering. Quantitative changes in the level of Flowering Locus C protein cause quantitative changes in flowering time. Low levels of Flowering Locus C protein cause the plant to flower earlier than high levels of Flowering Locus C protein. The *Flowering Locus C* gene integrates a huge amount of information flowing in from many genes of the vernalization and autonomous pathways.

Following the pathworks logic, we might predict that evolutionary changes in vernalization would most likely involve mutations that alter regulation of *Flowering Locus C* transcription. We might expect that the mutations with the least pleiotropic effects would occur in the *cis*-regulatory region of the *Flowering Locus C* gene itself. There are indeed two known mutations of the *Flowering Locus C* gene in *Arabidopsis thaliana* populations that alter vernalization. Both mutations alter the *cis*-regulatory region of the *Flowering Locus C* gene. So far, so good. But, as discussed repeatedly in this book, studies of natural populations suggest that the *Frigida* gene is a hot spot for evolutionary changes in vernalization. More than 20 independent null mutations in the *Frigida* gene alter the requirement for

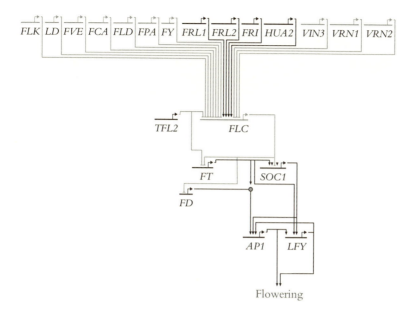

FIGURE 7.12. This figure illustrates the vernalization pathwork in *Arabidopsis thaliana*. The vernalization pathwork clarifies that all of the vernalization signals shown are integrated by the *Flowering Locus C* gene (*FLC*). The FLC protein represses several genes that would otherwise promote flowering. The Frigida (FRI) protein promotes transcription of the *FLC* gene, repressing flowering. Vernalization causes activation of multiple genes that repress transcription of the *FLC* gene, promoting flowering. Removal of the FRI protein upsets this quantitative balance of factors that regulate transcription of the *FLC* gene, resulting in lower levels of FLC protein and faster flowering.

vernalization in natural populations of *Arabidopsis thaliana*. Null mutations of the *Frigida* gene do alter transcription of the *Flowering Locus C* gene, but the *Frigida* gene does not sit in an obviously central control position for vernalization. The *Frigida* gene does not appear to be located in an optimal pathwork location for the accumulation of mutations with limited pleiotropic effects. In fact, null mutations of the *Frigida* gene do have pleiotropic effects on growth rate.

Why, then, has the *Frigida* gene accumulated so many null mutations in natural populations that alter the vernalization response? One possible explanation is that this pattern has resulted from the distribution of random mutations together with the effects of population structure and ecology. Null mutations are easily generated; many possible mutations can generate a null allele of the *Frigida* gene. Small, recent subpopulations of *Arabidopsis thaliana* that have experienced selection for loss of vernalization may have rapidly accumulated null mutations in the *Frigida* gene. In contrast, there are likely to be few mutations in the *Flowering Locus C* gene that can cause specific effects on *Flowering Locus C* transcription. Null mutations in the *Flowering Locus C* gene, which probably also arise frequently, would be unlikely to improve fitness because the *Flowering Locus C* gene integrates multiple signals to regulate flowering time. Null mutations in the *Flowering Locus C* gene would likely have excessive pleiotropic effects. It is possible, therefore, that the abundance of null mutations in the *Frigida* gene in natural populations of *Arabidopsis thaliana* reflects the combined effects of pathwork structure and population structure. The pathwork demands that mutations should somehow alter activity of the *Flowering Locus C* gene. Null mutations in the *Frigida* gene will do this. But the population structure of *Arabidopsis thaliana* may limit the pool of available mutations and these populations may be making the best of a bad situation by evolving via null mutations of the *Frigida* gene rather than with more specific mutations in the *cis*-regulatory region of the *Flowering Locus C* gene.

This last example implies that a deeper understanding of genetic evolution requires a synthesis of developmental biology and evolutionary biology. In the next chapter, I examine the specific claim that population processes matter for genetic evolution.

## SUMMARY

Parallel evolution provides evidence that some genes are hot spots for evolutionary change. Hot spots may result from the architecture of genetic networks. Regulatory networks often possess genes that act as central control points, which integrate diverse signals to generate a simple output, and these are likely to be evolutionary hot spots. For multicellular organisms, viewing gene-regulatory networks as pathworks, and tracing these pathworks backwards in time from the perspective of a single differentiated cell, clarifies the evolutionarily relevant aspects of network structure. Locations in pathworks in which mutations can maximize the fit between phenotypic change and selective regime, with minimal pleiotropic effects, are likely to be hot spots of genetic evolution. Development biases genetic evolution towards hot-spot genes. Pathworks alone cannot predict the distribution of mutations that contribute to evolutionary variation. Population structure may cause failure of pathwork predictions, perhaps, in part, by limiting the number of mutations available to be acted upon by selection.

# EIGHT

# THE PREDICTABLE GENOME

> When we look to the individuals of the same
> variety or sub-variety of our older cultivated plants
> and animals, one of the first points which strikes
> us, is, that they generally differ more from each
> other, than do the individuals of any one species or
> variety in a state of nature.
>
> —Charles Darwin, first sentence of *The Origin of Species*

We have seen that the phenotypic effects of mutations depend on where mutations occur within genes and within pathworks. Within genes, mutations tend to have recessive effects if they simply alter expression levels or quantitatively change catalytic activity. Mutations may have dominant effects if they generate a new expression pattern or if they generate a qualitatively new function. Within pathworks, genes located at integrative positions can accumulate mutations with specific effects—and these mutations will tend to occur in *cis*-regulatory regions—whereas mutations in genes upstream of integrative positions will tend to have pleiotropic effects. It is currently more difficult to predict which mutations will have epistatic effects. When many mutations arise in a population, selection will tend to favor mutations with stronger dominance and with weaker pleiotropic and epistatic effects. Thus, to the extent that genetic evolution is sensitive to dominance, pleiotropy, and epistasis, substitutions will occur preferentially in specific, and predictable, locations in pathworks and in genes.

Our ability to predict genome evolution is currently limited by our incomplete knowledge of gene function, of gene structure, and of pathworks. Current research is rapidly increasing our understanding of gene function and structure. A deep understanding of pathworks will take a little longer. This is because single genes often participate in many—perhaps hundreds or thousands—of pathworks. The contribution of a gene to each pathwork depends on specific *cis*-regulatory regions associated with that gene. For example, even though a pathworks view leads to the prediction that evolution of larval trichome patterns in *Drosophila* will often involve evolution of the *shavenbaby* gene, it would be silly—for at least two reasons—to look simply for an excess of substitutions along the entire length of the *shavenbaby* gene in species that have evolved new trichome patterns. First, only a small minority of the *shavenbaby* gene participates in the differentiation of the particular trichomes that have evolved. Some parts of the *shavenbaby* gene control development of trichomes that have not evolved, and other parts of the gene are involved in ovary development. The second reason that we cannot simply compare the substitution rate experienced by various genes in the trichome pathwork is that several genes in the pathwork may have evolved to generate other phenotypic changes in the organism. Predicting which genes have evolved requires knowledge of the pathwork responsible for the phenotype that has evolved. Making more detailed predictions requires an understanding of which parts of each gene participate in this particular pathwork.

Even if we had a complete knowledge of every functional element in the genome and of all pathworks, we still would be missing critical information that is required to make accurate predictions about genome evolution. This is because population size and structure probably influence genetic evolution. Small populations may be mutation-limited. Populations experiencing a low input of mutations in each generation may end up substituting alleles with pleiotropic and epistatic effects, simply because mutations with more specific effects have not appeared. In addition, small subpopulations of a single species may experience diverse selection pressures and may evolve using mutations that are not uniformly favorable throughout the species range.

If any of these hypotheses about the effect of population size and structure on genetic evolution are true, then predicting genetic evolution requires both a deep understanding of development and of the population context in which evolution has occurred. Considering both development and population genetics simultaneously complicates the task of predicting genetic evolution. It would be much easier if we could simply unravel development and then predict where in the genome evolution will occur. It would therefore be nice to know whether we really need to consider the population part of the equation. Do large populations really experience different kinds of genetic evolution than do small populations? Do large, undivided populations exposed to persistent selection experience different kinds of genetic evolution than do subdivided populations exposed to variable environmental conditions?

And, if any of these hypotheses is true, what are the long-term consequences for evolution? Data collected over the past two decades is beginning to provide clues to the answers.

⤳

It is useful to begin examining the data by looking at extreme evolutionary events, since dramatic examples often bring issues into sharp relief. For a sexual species, the most extreme form of natural selection is selection of two individuals—one male and one female—in each generation. This is similar to mutagenesis screens performed by geneticists. In a mutagenesis screen, a population of genetically homogeneous individuals is exposed to a mutagen that creates many DNA lesions. The geneticist then examines the progeny of the mutagenized population for individuals that carry a specific phenotypic defect. Usually, one individual displaying the desired defect is selected and mated to another individual without the defect. The causal mutation is then maintained in an inbred lineage.

This experiment and natural selection have the same logical structure. But a mutagenesis screen differs from selection in the wild in at least four ways. First, mutagenesis involves the strongest possible kind of selection. The fitness of the individual carrying the new mutation is 1 and the fitness of all other individuals is 0. Natural selection, by contrast, usually involves much weaker selection. As shown in Figure 8.1, selection on phenotypic traits in the wild shows an approximately exponential distribution, with some traits experiencing strong selection, but most traits experiencing selection closer to 0. Variation in most of these phenotypic traits probably results from the combined effects of multiple mutations, so the selection acting on each mutation is likely to be considerably smaller than the selection

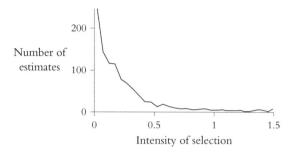

FIGURE 8.1. Estimates of natural selection on phenotypic traits in wild populations show an approximately exponential distribution. The intensity of selection is measured here as the selection gradient, which does not equal the selection coefficient that was encountered earlier in the book. See the Notes for a discussion of the selection gradient.

acting on each phenotypic trait. Second, geneticists usually select for phenotypic changes that are more extreme than those that get selected in wild populations. Third, alleles derived from a mutagenesis screen often are maintained in a single homogeneous genetic background, whereas alleles in natural populations are exposed to a new genetic background in every individual in every generation. Fourth, laboratory populations usually are maintained in relatively constant environments, so laboratory populations experience only weak selection for the ability to cope with a fluctuating environment.

If the strength and pattern of selection and the population history have any effect on the kinds of mutations that evolve in populations, then mutagenesis screens should provide examples of mutations that differ in some way from the mutations that are selected in natural populations. They do. Alleles isolated in mutagenesis screens often cause extensive pleiotropic effects. Many are homozygous lethal, which is an extreme form of deleterious pleiotropy. When these mutations are tested in a variety of more natural genetic backgrounds, these mutations often display epistasis with respect to natural variation. We observed examples of this epistasis in Chapter 5, in the form of hidden genetic variation, for mutations in the genes *Antennapedia* and *Ultrabithorax* that were introgressed into lines derived from wild *Drosophila melanogaster* populations. Thus, mutations resulting from mutagenesis screens, which have proven so useful in molecular and developmental biology, often show strong pleiotropic and epistatic effects.

Mutagenesis experiments often select for null alleles of evolutionarily conserved genes. This is particularly true for genes that are essential for core cell-biological processes and for genes that are involved in development. Null alleles in highly conserved genes have played a small role in long-term evolution, despite the ease with which null alleles arise. While some natural populations contain null alleles at many genes, the null alleles at each individual gene usually are found at low frequency. The low frequency in natural populations of null mutations at each gene suggests that these mutations are normally deleterious. Further, if null alleles in developmental genes played an important role in developmental evolution, then we would observe few highly conserved developmental genes. Mutations like those isolated in mutagenesis screens probably contribute little to long-term evolution. Mutagenesis experiments thus provide our first empirical clue that the strength of selection and population history influence genetic evolution.

Mutagenesis experiments constitute such extreme forms of evolution that it is unclear whether evolutionary conclusions drawn from the distribution of mutations found in mutagenesis experiments are relevant to any natural population. Our ideal data set would consist of observations of many mutations generating phenotypic evolution from populations of different sizes, with different population structures and experiencing different strengths, durations, and patterns of selection. Ideally, we would like to compare these mutations with the mutations generating differences between species for lineages that have experienced various—but known—population parameters and selective histories. The reader should have no difficulty seeing how hard it will be to collect such a data set. While evolutionary genetics has seen enormous progress in recent years—with the identification of many mutations contributing to phenotypic variation—we are still far from generating the kind of unbiased data set that is required to test models of genetic evolution.

More than 300 mutations causing phenotypic variation in domesticated races, within species, and between species have been reported over the past several decades. Many of these mutations were identified by studies that focused on candidate genes, and this fact, along with many others that I discuss at length in the Notes section at the end of this book, render the body of existing data a highly suspect substrate for testing models of genetic evolution. The existing data do, nonetheless, reveal unanticipated trends, and these trends provide a glimpse of the possible effects of ecological and population parameters on genetic evolution. At the very least, these trends provide a strong impetus to collect more data.

Because these data were collected largely using biased approaches, I avoid drawing conclusions that would result obviously from the known biases. For example, this survey revealed 234 mutations altering coding regions and 74 mutations altering *cis*-regulatory regions, but we cannot draw any conclusion from this apparent excess of coding changes since coding changes are far easier to discover than are *cis*-regulatory changes. In contrast, it is unlikely that the data collected at different taxonomic levels would reflect investigator or methodological bias. For this reason, I compare the kinds of mutations that cause phenotypic differences at different taxonomic levels. For example, we can divide the data set into mutations that cause characteristics of domesticated races, mutations that cause variation within a species, and mutations that cause differences between species. I will discuss differences between domesticated races and variation within natural populations in more detail later, but, for now, it is sufficient to recognize that human selection for domesticated races is likely to impose different population structures and selective regimes than those experienced by natural populations. Under the classical model of genetic evolution, we expect that variation within wild populations will probably resemble variation between species. If we first examine a broad category of mutations, null mutations, we do not see this expected pattern. Instead, the variation that we see within natural populations more closely resembles variation seen in domesticated races. As shown in Table 8.1, null mutations frequently cause phenotypic variation both within populations (25%) and in domesticated races (56%), whereas only 7% of the mutations causing species differences are null mutations.

Evolution leading to differences between species appears to be biased against null mutations. Why might this be true? The prevalence of null mutations of the *Frigida* gene in populations of *Arabidopsis thaliana* suggests a possible answer. Populations of *Arabidopsis thaliana* became established in new geographical locations as Europeans migrated across the globe. The prevalence of null mutations of the *Frigida* gene likely resulted from recent strong selection for rapid flowering in isolated popula-

## TABLE 8.1

Distribution of mutations causing phenotypic variation
arranged by taxonomic level.

|  | Domesticated | Variation within species | Differences between species |
|---|---|---|---|
| Coding | 65 | 122 | 47 |
| Cis-regulatory | 23 | 24 | 27 |
| Other | 11 | 11 | 1 |
| Total | 99 | 157 | 75 |
| Null | 55 | 39 | 5 |

tions. *Arabidopsis thaliana* populations are therefore similar to domesticated populations, which also have experienced recent strong selection and have accumulated many null mutations.

Like null mutations, *cis*-regulatory mutations that involve insertion or deletion of large pieces of DNA occur often within species, but they rarely lead to species differences. For example, 24 *cis*-regulatory mutations have been found that cause intraspecific phenotypic variation. In seven of these cases (29%), the mutations involve insertions or deletions of large pieces of DNA in the *cis*-regulatory regions of genes, altering normal gene expression. For example, a transposable element insertion upstream from a gene encoding a Cytochrome P450 protein in *Drosophila melanogaster* causes increased gene expression in specific tissues and confers insecticide resistance. In contrast, large insertion or deletion events are implicated in zero out of 27 cases of *cis*-regulatory evolution between species.

Within species, even subtle changes in phenotype often are caused by disruptive mutations. For example, in a sample of wild flies of *Drosophila melanogaster*, flies carrying large insertions in a 106,000 bp region encompassing the *achaete* and *scute* genes had an average of 1.62 fewer sternopleural bristles and 1.18 fewer abdominal bristles than did flies without these insertions. These insertions account for about 5% of the total variance in bristle number in this sample of flies. Most of the insertions are rare—usually singletons in population surveys—implying that they are deleterious and will soon be eliminated by selection. One of these insertions, however, was found at a frequency of 14%, suggesting that it is not universally deleterious and may even be maintained in the population by balancing selection.

These are extreme examples of mutations that generate variation within species but that rarely lead to differences between species. A more fine-grained analysis reveals that many less disruptive mutations show a similar pattern. Within species, about 20% of the mutations causing morphological evolution occur in *cis*-regulatory regions. But between species, as shown in Figure 8.2, about 80% of the mutations causing morphological evolution occur in *cis*-regulatory regions. The excess of *cis*-regulatory mutations causing morphological evolution between species is consistent with theoretical arguments discussed in Chapter 4: mutations in coding regions probably cause more pleiotropic effects than do mutations in *cis*-regulatory regions, particularly for genes involved in developmental patterning. We expect that morphological evolution will more often be caused by *cis*-regulatory mutations than by coding mutations. However, as shown in Figure 8.2, neither within species nor in domesticated races do we observe an excess of *cis*-regulatory mutations causing morphological evolution. These populations do not appear to follow the simple predictions regarding pleiotropy laid out in Chapter 4.

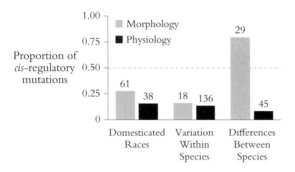

FIGURE 8.2. The proportion of evolutionarily relevant mutations occurring in *cis*-regulatory regions for different taxonomic levels. Mutations generating domesticated species and variation within species occur mainly in coding regions for both morphological and physiological evolution. Morphological differences between species, however, are caused mainly by mutations in *cis*-regulatory regions. The total number of mutations in each category are shown above the bars.

For physiological evolution, similar proportions of mutations occur in *cis*-regulatory regions, both within species and between species, as shown in Figure 8.2. This observation suggests that morphological evolution and physiological evolution follow different rules. It is not entirely clear why evolution of developmental and physiological genes would differ in this way. One hypothesis is that coding mutations in developmental genes have more pleiotropic effects than do coding mutations in physiological genes. This pattern would result, for example, if developmental genes were typically embedded in the middle of pathworks, whereas physiological genes were usually positioned towards the ends of pathworks. Following this logic, genes embedded in the middle of pathworks would be expected to possess large *cis*-regulatory regions that determine complex patterns of gene expression, whereas genes located at the ends of pathworks would likely have simpler *cis*-regulatory regions. Developmental genes may, therefore, contain more *cis*-regulatory targets for adaptive mutations.

∽

This brief survey of mutations contributing to phenotypic evolution yields up tantalizing hints that population parameters influence genetic evolution. The three categories offered—domestication, variation within species, and differences between species—greatly oversimplify the diversity of population structures and selective regimes found in nature. We can expand our discussion to additional categories to explore how evolutionary variables might influence genetic evolution. The schema presented in Figure 8.3 provides an initial framework for thinking about this problem.

The geometric shapes in Figure 8.3 represent the relative magnitude of variables shown along the top of the figure in the populations shown along the left. For example, populations exposed to mutagenesis screens usually have a small effective population size and experience strong selection. I have divided the effective population size into two components, the historical and the contemporary effective population sizes. The historical effective population size reflects the population size over past generations and influences the amount of variation present in the population. The contemporary effective population size reflects the number of breeding individuals actually exposed to selection during the most recent generation. It is possible for a population to have experienced a recent bottleneck—and thus for it to contain little genetic variation—and for the population now to have a large size that allows efficient natural selection.

160

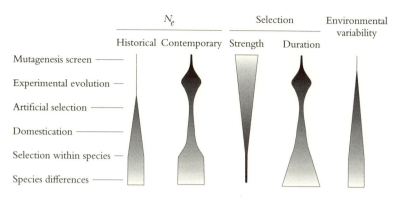

FIGURE 8.3. A schema for considering the effects of population size, selection, and environmental variability on the mutations fixed by natural selection. Different kinds of mutations may be favored in different kinds of populations, shown on the left. Artificial selection, domestication, and selection within species are ranked as examples grading from populations with small effective population sizes ($N_e$) to large, strong selection to weak, and relatively constant environments to variable environments. Many specific populations will defy this ranking. For example, some natural populations probably have smaller effective population sizes than some domesticated races.

Figure 8.3 is a heuristic and should not be interpreted literally. Natural populations experience a diversity of ecological conditions that cannot be divided neatly into a small number of categories. In addition, the variables in this schema should be interpreted with care. For example, the duration of selection implicitly incorporates other processes, since other events may occur during a long episode of selection. Local adaptation in subdivided populations may prevail during a short episode of selection, whereas selection over a long period of time may result in evolution of universally advantageous alleles.

The schema in Figure 8.3 includes separate lines for experimental evolution and artificial selection, even though many authors conflate these terms. Similarly, domestication is sometimes described as artificial selection. I define these three terms more narrowly to highlight several critical differences between them.

During experimental evolution, a population is introduced into a novel environment and evolves higher fitness. Experimental evolution tends to be performed on microorganisms—usually viruses, bacteria, or yeasts—that reproduce asexually during the experiment. Typically, the experiment is initiated with a single clone, so there is initially no (or very little) variation in the population. As the population grows quickly to a large size, usually to greater than $10^9$, it experiences selection for optimal growth under the novel environmental conditions. Many new mutations are introduced into the population every generation, since the contemporary effective population size is large. New adaptive mutations often sweep through the population, and adaptive evolution usually involves a series of consecutive selective sweeps. Consecutive substitutions may show epistasis with respect to one other. This experimental design probably replicates, or nearly replicates, how many microorganisms actually evolve in the wild. When a microorganism first enters a new ecological niche, like a human body, the population size grows rapidly, new mutations are introduced, and selection in the novel environment causes rapid serial evolution of new mutations.

Experimental evolution closely resembles mutagenesis experiments, except that, in experimental evolution, the contemporary effective population size is huge and experiments are performed over many generations. These two facts mean that many new mutations will be introduced and available for selection during the course of an experiment. Also, recessive lethal mutations, which are commonly found in mutagenesis screens, will not substitute in an experimental-evolution population.

Replicate experimental-evolution populations often evolve similar or identical genetic changes. For example, when the bacteriophage $\phi$X 174 was reared at a high temperature in two replicate populations in a novel host, *Salmonella typhimurium*, half of the same codons mutated to the same amino acid in both replicates. Repeated selection of the same mutations supports the hypothesis that these particular mutations contributed to increased bacteriophage fitness when grown in the novel host. This hypothesis is further supported by the observation that four out of 19 of the substitutions selected during experimental evolution were identical to the amino-acid substitutions found in a closely related bacteriophage that was isolated originally from *Salmonella*. Thus, experimental evolution provides compelling evidence for parallel evolution at the nucleotide level.

Artificial selection involves human-guided selection in each generation, and it differs from experimental evolution in four critical ways. First, the starting population in artificial selection consists of multiple genetically diverse individuals, since one goal is to determine how the variation present in a sample of individuals taken from nature responds to a particular selective regime. Second, artificial selection experiments are normally performed for fewer than 100 generations. Historically, this was done to minimize the contribution of new mutations to the experiment. It turns out, however, that new mutations contribute to an increasing fraction of the response to selection after about the first 20 generations. Thus, evolution in artificial selection experiments that persist for longer than 20 generations results both from selection of the original variation present in the population and from selection of new mutations introduced during the experiment. Third, normally artificial selection is performed on sexually reproducing species. Thus, selected mutations experience a new genetic background in each individual and may have epistatic fitness effects in these different genetic backgrounds. Fourth, often artificial selection involves selecting a small fraction of the population for breeding. Thus, the effective population size is small, and only alleles of relatively large effect can be selected. However, the relatively small effective population size allows some mutations with weak effects on fitness to spread by drift.

Compared with experimental evolution, artificial selection involves smaller contemporary effective population sizes and fewer generations of selection. In addition, the experiments start with standing variation, so the response to selection in the first tens of generations results from selection of variation sampled from nature. By and large, this variation will have been either neutral or under balancing selection and likely to have been either formerly segregating at intermediate frequency or deleterious and present in the original population at low frequency.

Artificial selection experiments result often in selection of variants with epistatic and perhaps pleiotropic effects. As shown in Figure 8.4, selection for increased and decreased numbers of bristles on one part of the abdomen resulted in a strong evolutionary change in abdominal bristle number. This evolutionary response involved selection of alleles at multiple loci. Many of these alleles also changed the number of bristles on the thorax, suggesting that these alleles had pleiotropic effects. In addition, many of these alleles showed large epistatic effects with respect to each other, and the evolved alleles at two loci were associated with a significant reduction in the viability

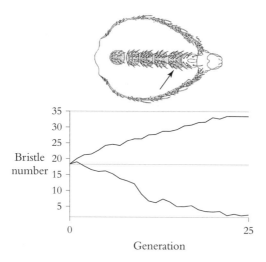

FIGURE 8.4. Artificial selection on abdominal bristle number. The parental population, which was a mixture of 62 isofemale lines, displayed a mean number of about 18 bristles. Two lines were selected over 25 generations, one for increasing bristle number and one for decreasing bristle number. Each generation, the top (or bottom) 25 males and 25 females were selected from 100 flies of each sex that were measured. The number of bristles was counted on abdominal sternite 6 of females (*arrow*) and sternite 5 of males.

of the fly. Similar results have been observed in many artificial-selection experiments when selection has been performed for fewer than about 50 generations. This suggests that variation commonly found in natural populations exhibits extensive epistasis and possibly pleiotropy.

When artificial selection has been performed for more than 50 generations, the responses to selection in later generations appear to have resulted largely from new mutations that arose during the experiment. In one example, selection on abdominal bristle numbers for more than 80 generations led to a dramatic increase in bristle number in six replicate lines. The increases in bristle number in later generations resulted largely from selection for recessive lethal mutations and for visible mutations of large effect that arose during the experiment. The mutations selected during these later stages are apparently similar to mutations selected during experimental evolution and during mutagenesis experiments.

Ever since Darwin introduced the idea, domestication has been held as a model of how evolution occurs in wild populations. Domestication is a complex process, involving both historical selection over a long period of time, which produced the general characteristics of domesticated species, and more recent selection, which generated particular varieties. Some authors call the first step "domestication" and the subsequent production of varieties "improvement". It is likely that domestication involved different population sizes, population structures, and strengths of selection than did improvement. It may therefore be instructive to divide traits of domesticated species into those that were likely generated during early domestication and those reflecting later improvement. We have a better understanding of the process of improvement than of historical domestication, simply because most improvement occurred during historically recent times, whereas domestication of most crops and animals occurred during prehistory. Unfortunately, few mutations contributing to historical domestication have been identified, so it is not yet possible to perform a proper comparative analysis of domestication and improvement. However, a brief review of several cases reveals that improvement often involves mutations that strongly disrupt gene function, whereas domestication seems to involve mutations with more subtle effects on gene function.

FIGURE 8.5. A Belgian Blue bull homozygous for a presumptive null allele of *myostatin*.

Many improvement traits result from null mutations or other mutations of large effect. The mutation that causes wrinkled pea seeds, first studied by Gregor Mendel, is a transposable-element insertion that inactivates a gene producing a starch-branching enzyme. Six different null mutations in the *myostatin* gene cause muscle hypertrophy in different breeds of cattle, an example of which is shown in Figure 8.5. A null mutation in the *myostatin* gene also causes muscle hypertrophy in some individuals of the whippet dog breed. The *myostatin* gene encodes a protein similar to many other proteins known as the Transforming Growth Factor-$\beta$ superfamily of growth factors and acts as a negative regulator of muscle development. Homologs of this gene are found in all mammals. While null mutations of the *myostatin* gene result in cattle with more and leaner meat, the cows have difficulty calving. Thus, null mutations in the *myostatin* gene cause negative pleiotropic effects.

Some phenotypic variation in domesticated races is caused by altered gene regulation and not by null mutations, but even in these cases the mutations have dramatic effects on gene function. White grapes, for example, are caused by insertion of a transposable element upstream of the *VvmybA1* gene, which encodes a transcription factor that regulates anthocyanin biosynthesis. This mutation and many others contributing to improvement are reminiscent of many of the mutations discovered during the early decades of genetics research on *Drosophila melanogaster*, which were generated by insertion of transposable elements into or near genes. The strong selection on domesticated varieties apparently has been sufficient to overcome deleterious pleiotropic effects of these mutations.

The mutations discovered so far that contributed to historical domestication cause more subtle changes in gene function than do the improvement mutations. For example, domesticated pigs have more muscle mass than do wild boars. A single base pair change in the *cis*-regulatory region of the *insulin-like growth factor 2* gene controls 15–30% of the difference in muscle mass.

Similarly, some of the mutations causing changes in the overall architecture of maize also have been localized to the *cis*-regulatory region of a gene. Teosinte plants normally exhibit multiple stalks, each with some small lateral ears and a terminal tassel, as shown in Figure 8.6. In contrast, maize typically produces a single strong stalk with several lateral ears and a terminal tassel. This fundamental difference in plant architecture was caused almost entirely by evolution of the *teosinte branched1* gene, which encodes a transcription factor. The maize allele

Teosinte          Maize

FIGURE 8.6. Differences in the plant architecture of teosinte (on the left) and maize (on the right). Teosinte produces multiple axial branches each bearing small ears. Maize produces a single main stem with several large ears.

produces twice as much Teosinte branched1 transcript as is produced by the teosinte allele. The higher levels of Teosinte branched1 transcript result from one or more mutations in a *cis*-regulatory region far upstream of the first exon of the *teosinte branched1* gene.

In summary, examples of improvement seem similar to many cases of variation observed within species and to the results of mutagenesis experiments, with many null mutations contributing to improvement. The few mutations known that have contributed to historical domestication seem similar to differences observed between species, with an excess of *cis*-regulatory mutations.

Evidence from laboratory experiments and from evolution in the wild appears to be consistent with the hypothesis that population history, population structure, and the strength of selection influence genetic evolution. More precisely, large populations that have experienced selection over long periods of time tend to evolve via mutations with specific effects. The most important difficulty with this idea is that predictions from population genetics are notoriously dependent on multiple parameters, each of which is difficult to measure. As discussed in Chapter 3, even apparently simple predictions about the contribution of dominance to evolution are confounded by whether the relevant mutations are or are not present at mutation–selection equilibrium. Most of the intuitive explanations of population genetics that I have provided in this book are based on classical population genetics theory, which typically includes many assumptions that may not be valid in every case. Amongst the assumptions most likely to be suspect and problematic are assumptions of constant selection coefficients and of constant population sizes. Similarly, the pathworks predictions outlined in Chapter 7 are based on fairly simple and nonquantitative descriptions of gene interactions. It is not clear whether the structure of pathworks will remain so simple—with, for example, the presence of central integrating genes—when pathworks are studied in more detail. In addition, it is not clear whether these nonquantitative descriptions of pathworks will remain valid when quantitative measurements are included.

A second major difficulty is that we really have little idea of the relative quantity of new mutations with specific effects versus those with pleiotropic and epistatic effects. Mutations with specific effects may simply be too rare to contribute often to evolution. Evolution may have to make due with suboptimal mutations, simply because these are all that can be generated on a reasonable time scale.

A further difficulty with these ideas is that any single observation is subject to post hoc explanation. Darwin famously wrote, in *The Origin of Species*, "If it could be demonstrated that any complex organ existed, which could not possibly have been formed by numerous, successive, slight modifications, my theory would absolutely break down." Unfortunately, predictions from the integration of pathwork thinking and population genetics are not so easily rejected. Imagine that you discover a mutation that contributes to an evolutionary change and that this mutation alters the function of a gene that happens to sit in a central, integrative position of a pathwork. Certainly, you might argue that this is what we would expect, given the structure of the pathwork. Likewise, if the mutation alters a gene that resides somewhere else in the pathwork, it would be easy to argue that unusual population structure has resulted in selection of a nonoptimal mutation. This is the temptation when presented with any optimality model: claim success when the data fit and invoke other forces when the data do not fit. On the other hand, if we fail to incorporate population thinking, we may end up rejecting pathworks predictions for the wrong reasons.

One way to rigorously test these ideas is to generate a large body of data on the mutations causing phenotypic variation and to explore these data in the context of independent, quantitative measures of population parameters. Single case histories—such as those I have exploited in this chapter to outline the argument—can neither rigorously support nor ultimately disprove the idea that pathworks and population history interact to make genetic evolution predictable.

In the best of conditions, historical processes can often be explained by multiple competing hypotheses. In this book, I have asked whether developmental biology can be combined with population genetics to explain any patterns of genetic evolution. The conceptual framework provided by deconstructing development as pathworks may help us to reveal these patterns. But predictions from pathworks thinking will hold only when populations can test a large number of mutations to select those with the most favorable effects. Selection in small populations and selection over a short period of time will sample only a small subset of possible mutations. Therefore, I have asked whether different population parameters lead to different modes of genetic evolution. If this is true, then we need to synthesize population genetics with pathwork thinking to generate relevant predictions about genetic evolution. Genetic evolution in small populations is not necessarily unpredictable, just different from evolution in large populations. Evolution in small populations may often involve selection on mutations in the same genes—think of the many null mutations of the *Frigida* gene in *Arabidopsis thaliana*—and, as we learn more about pathworks, it may become clearer why these mutations have been favored. But we should not expect genetic evolution inevitably to proceed in similar ways in small and large populations.

Given the abundance of mutations in natural populations that do not appear to follow predictions derived from pathworks thinking, such as the multiple null mutations contributing to phenotypic variation in domesticated races and within species, it is reasonable to ask whether these mutations represent evolutionary dead ends in isolated populations or whether they play any role in contributing to long-term evolution. The existing data are far from sufficient to answer this question, but we can speculate.

While I predict that large populations will most often evolve via mutations that disrupt development as little as possible—for example, by mutations that hit genes in integrative positions in pathworks—small populations may sometimes evolve via mutations that strongly disrupt pathworks. As we saw in Chapter 3, small populations may favor the evolution of dramatic genome changes, such as gene duplications. Small populations also are likely to experience dramatic changes in population size and strong and fluctuating natural selection, as we saw for finches on the Galapagos Islands. Do these conditions, which may lead to evolution by fixation of nonoptimal mutations, doom these populations? Are mutations that disrupt pathworks toxic to the survival of species? In some cases, this might be true. But there is a flip side, a creative side, to these mutations. Mutations that follow pathworks predictions are the well-behaved, nondisruptive mutations of the genome: they don't ruffle feathers, they don't disrupt genetic networks. These mutations often alter linkages in genetic networks—think of all the *cis*-regulatory mutations contributing to phenotypic evolution—but they alter network linkages in specific ways that minimize pleiotropic effects. It is not yet clear whether the accumulation of these specific changes is sufficient to generate the extensive rearrangements to genetic networks that have generated the diversity of life on Earth. Mutations that disrupt pathworks offer an interesting contrast. These mutations— mutations such as gene duplications and other dramatic changes to gene function—have the capacity to dramatically alter regulatory connections. Such mutations have little chance of spreading in large populations experiencing efficient purifying selection. Evolution via disruptive mutations is more likely in small populations. Perhaps—just perhaps—while most populations evolve obediently through the fixation of mutations with specific effects, small populations, while teetering toward extinction and irrelevance, provide cauldrons of evolutionary novelty.

## SUMMARY

The probability that a new adaptive mutation is fixed in a population depends on how the mutation alters development to generate a change in the phenotype. Mutations expressing stronger dominance and with fewer pleiotropic and epistatic effects are more likely to be substituted than are mutations demonstrating weaker dominance and more extensive pleiotropic and epistatic effects. These predictions rely on natural selection sampling repeatedly from a large pool of mutations with different kinds of effects. Genetic evolution in small populations, therefore, may not follow these predictions. Current data suggest that population parameters have influenced genetic evolution. Species with large, panmictic populations and that have experienced persistent, moderate selection appear often to evolve via mutations with specific effects. Species with small population sizes, or with highly structured populations, or that have experienced intense, brief episodes of selection have often evolved via mutations with more dramatic effects. Given knowledge of pathworks and population parameters, genetic evolution appears to be predictable.

# THE HOPEFUL MONSTER IS DEAD! LONG LIVE THE HOT SPOT!

The existence of genetic hot spots may resolve an old debate within evolutionary biology. Ever since Darwin emphasized in *The Origin of Species* that evolutionary changes involve small steps, other scientists have countered that large steps seem reasonable. This debate heated up toward the end of the nineteenth century, when William Bateson emphasized that evolution by small steps—what we can call micromutations— seemed unable to explain the apparent phenotypic gaps between higher taxa. Instead, it seemed more likely to him that mutations of large effect, so called macromutations, provided a better explanation for dramatic differences between higher taxa. Bateson emphasized the existence of mutations causing systemic changes to the phenotype. In particular, he discovered many examples in nature of what he called homeosis, or the transformation of one organ into the likeness of another organ, as shown in Figure HN.1. Indeed, these mutations seem to recapitulate inferred evolutionary transitions between higher taxa in a single step!

FIGURE HN.1. Drawing of the head of a specimen of the lobster *Palinurus penicillatus* possessing a homeotic transformation of the left eye into an antenna.

These observations led Bateson and others, such as Thomas Hunt Morgan, to emphasize the instantaneous creative power of mutation over the plodding workings of natural selection on small variations. As compelling as these observations seem, in the early years of the twentieth century evolutionary biologists recognized that this model was incompatible with observations of the fine phenotypic gradations within and between closely related species. In addition, evolutionary biologists constructed mathematical models to show that natural selection on Mendelian mutations of small effect was more than sufficient to explain genetic diversity at all levels, given the vast amounts of time involved.

Richard Goldschmidt briefly reignited this debate in 1940 by publishing *The Material Basis of Evolution*. Goldschmidt argued that the mutations causing intraspecific variation do not contribute to differences between species. Instead, he argued that mutations of more dramatic effect, mutations that generated what he called hopeful monsters, caused species differences. This book was rapidly dismissed by many leading evolutionary biologists, and it has served as the whipping post of macromutationism ever since.

Participants in this debate often overlooked the fact that both sides were motivated by observations of natural populations and of induced mutations. Here are the central facts gleaned from observations of natural populations. First, most populations harbor extensive reserves of genetic variation consisting mostly of mutations of small effect. Second, closely related species tend to differ in small and quantitative ways from each other. Third, progressively more distantly related species tend to look increasingly different from one another. Fourth, distantly related species display obvious discontinuities—jumps in phenotypic space. This last pattern must result either from extinction of intermediate forms or from mutations that cause discontinuities.

The central facts gleaned from observations of induced mutations are these. First, many mutations cause quantitative changes in shape, size, or number of body parts and subtle changes in behavior and physiology. Second, many mutations cause discrete changes in the phenotype, such as changing eye color from red to white. Micromutationists have tended to emphasize the importance of the first class of mutations. Macromutationists have emphasized the second class of mutations and particularly those mutations that resemble observed evolutionary transitions, only as if viewed backwards in time. For example, a set of mutations at the *Ultrabithorax* gene in *Drosophila melanogaster* cause homeotic transformation of the third thoracic segment into the likeness of the second thoracic segment, as shown in Figure 4.3. These flies possess a pair of wings in place of the halteres that normally reside on the third thoracic segment. This arrangement of wings superficially resembles the ancestral state, which is illustrated by butterflies and dragonflies in Figure HN.2.

FIGURE HN.2. Homeosis only superficially resembles evolutionary transitions. The "bithorax" fly on the left is caused by several mutations that eliminate *Ultrabithorax* activity in the third thoracic segment. The halteres on the third thoracic segment are transformed into wings that are virtually indistinguishable from wings on the second thoracic segment. The butterfly in the center (*Idea idea*) and the dragonfly on the right *(Epiaschna heros)* illustrate that the hind wings of other insects are different from the forewings.

Macromutationism has been criticized for many reasons, but there are two critical problems with it. First, mutations of dramatic effect only *resemble* evolutionary transitions. They do not precisely *match* evolutionary transitions. For example, the transformed hind wings of the *Ultrabithorax* mutant fly look identical to the forewings. In contrast, during evolution, the haltere evolved from a hind wing that looked different from the forewing. In all known and fossil insects, the hind wings look different from the forewings, as can be seen in the butterfly and dragonfly shown in Figure HN.2. This is because, in all insects examined so far, expression of Ultrabithorax protein in the hind wings causes the hind wings to look different from the forewings. Dramatic change in the expression pattern of the Ultrabithorax protein is, therefore, an unlikely route to the evolution of halteres. Darwin recognized this general difficulty.

> Under domestication monstrosities sometimes occur which resemble normal structures in widely different animals. Thus pigs have occasionally been born with a sort of proboscis, and if any wild species of the same genus had naturally possessed a proboscis, it might have been

argued that this had appeared as a monstrosity; but I have as yet failed
to find, after diligent search, cases of monstrosities resembling normal
structures in nearly allied forms, and these alone bear on the question.
[*The Origin of Species*, pp. 38–39]

The second problem with macromutationism is that mutations of large phenotypic effect tend to generate deleterious
pleiotropic effects. Mutations that cause extensive phenotypic
alterations, such as homeotic mutations, usually involve altered
expression of transcription factors that perform many roles
during development. Thus, dramatic changes in expression of
these genes often cause pleiotropic effects.

Macromutationism made a slight comeback with the rise of
evolutionary developmental biology. Starting in the 1980s, evolutionary developmental biologists examined the evolution of
so-called homeotic genes—genes that can be mutated to cause
homeosis. Often, the expression patterns of these genes correlate with dramatic differences in body plans between distantly
related taxa. This seems to support the idea that dramatic mutations in these genes contributed to evolution. However, this
argument is hamstrung by the likely pleiotropic effects of such
mutations. Hot spots provide a solution to this conundrum.

If mutations affecting a single trait tend to occur at a limited
number of hot-spot genes, then, over time, these genes will
accumulate many mutations and may ultimately end up with
dramatically different functions in distantly related species. This
hypothesis is consistent with all observations and theory from a
micromutationist view of evolution: the individual mutations
generating variation within species and between species tend to
have small effects. This hypothesis is also consistent with observations across large taxonomic distances; divergent species
sometimes possess homologous genes with strongly divergent

functions. The *shavenbaby* gene, for example, functions in different ways in *Drosophila melanogaster* and in *Drosophila sechellia*, but these differences were generated by the accumulation of multiple mutations of smaller effect. Micromutationism and macromutationism reflect observations of the same phenomena at different temporal scales. Evolution via the accumulation of mutations at hot-spot genes is still evolution by relatively small steps, but not in the way originally envisioned by most evolutionary biologists over the past century. In the traditional view, evolutionary biologists expected to observe mutations of small effect scattered across hundreds or thousands of different genes. Instead, many of the mutations contributing to evolution may be concentrated at hot-spot genes.

# ACKNOWLEDGEMENTS

If I had known before I started the resistant path I had to travel to finish this book, then this book would likely not exist. The current book has not simply evolved from revisions of earlier versions; it has resulted from a wholesale revision of my way of thinking about the problem and, more concretely, from discarding preconceptions and assumptions that I had acquired during nearly twenty years in science. I could not have made this journey without the intellectual contributions of many people over the past five years.

First, I thank John Bonner for reading an early draft of the first few chapters of the very first version. While almost nothing from that draft made it into the final book, his enthusiastic endorsement of the project ensured that I persisted past my initial tentative trials. Greg Gibson encouraged me to write *any* book, even before I had thought of this one. Early on, Simon Levin, Steve Pacala, Dolph Schluter, and John Gillespie (perhaps unwittingly) provided critical insights during casual discussions. I thank Enrico Coen, Gregory Davis, Bronwyn Duffy, Marcus Feldman, Greg Gibson, Peter Grant, Rosemary Grant, Hopi Hoekstra, Virginie Orgogozo, Mark Ptashne, Margarita Womack, Matthew Rockman, Dolph Schluter, Saeed Tavazoie, John True, Adam Wilkins, and Patricia Wittkopp for remarkably kind (with a few exceptions) and thorough comments on the first draft. This version evoked widely divergent opinions, amounting, on average, to "nice try, but try again." After much soul searching and a good push from Ptashne, I started again with a blank page. Version two was definitely different, but, alas, it also got scrapped. I both thank and apologize to everyone who attended Enrico Coen's 2007 lab retreat and who tore apart every page of version two. Fortunately, large volumes of good wine blunted the verbal rapiers, and I emerged with version three. Two individuals, in particular, were critical to clarifying and sharpening my argument. First, Enrico Coen helpfully protested every time he felt I had used a concept sloppily or had used a sloppy concept; and, second, Virginie Orgogozo argued with essentially everything I wrote. May we all be blessed with such honest colleagues. I am grateful to Caroline Dean, Justin Gerke, Alistair McGregor, François Payre, John True, and Roman Yukilevich for critical reading of sections of the final version, and to Edmund Brodie III, Tim Cooper, and Joe Thornton for providing modified figures and raw data from their published work. Justin Gerke, François Payre, and Caroline Dean and David Laurie provided extraordinarily helpful advice on navigating the pathworks of yeast sporulation, *Drosophila* trichome determination, and *Arabidopsis* flowering time, respectively. A near-final draft was expertly dissected by Peter Andolfatto, Armand Leroi, Molly Przeworski, and Michael Whitlock. These reviewers provided so many helpful suggestions that I ended up going back and rewriting almost every chapter, some of them quite extensively. Both Peter Andolfatto and Michael Whitlock patiently reread multiple drafts of troublesome passages. The book was expertly copyedited by Gunder Hefta and proofread by Anna Liebowitz. Needless to say, all of these colleagues deserve enormous credit for any value in the book and none of the blame for any blemishes that remain.

ACKNOWLEDGEMENTS

I sometimes joke—to my publisher's dismay—that this book will have been read by more people before it was published than it will be afterward. It is entirely possible that I have forgotten one or two people who provided comments or helpful criticisms at some point on this project. For this, I apologize, and I thank the unacknowledged colleague.

I owe a deep debt of gratitude to all the members of my laboratory, past and present, who generated the data that has forced me to look at evolution with new eyes. It is largely by trying to make sense of our experimental results that I found myself on the path that led to this book.

Finally, I do not know who has suffered more: me, my wife, or my publisher. When I started this project, I didn't know what role publishers played in writing. I now know what role they can play, if they happen to be Ben Roberts. As should by now be clear, this book has been an odyssey. Ben was at my side throughout, and several times he talked me off the ledge.

The influence of my wife, Bronwyn Duffy, is here on every page. Her sharp ear for language (and her sharper tongue) drove me to revise almost every sentence. She copyedited with an unyielding pen and, by reading the book aloud to me, identified serious structural problems with earlier versions. Whatever clarity appears in these pages comes largely from her dedication to this project, which pretty much explains why I am forbidden to write another book until our son graduates from high school.

# NOTES

## PREFACE

**P. xi: Differences between individuals—the raw material of evolution— are caused by variation in genes and by variation in the environment.** More strictly, phenotypic variation results from genetic differences, environmental variation, and the response of genetic variants to environmental variation. This last component—the complex interplay of genetic and environmental variation—is often called genotype-by-environment interaction. The variance of a sample of phenotypic measurements can be decomposed into the variance caused by genetic differences, the variance caused by environmental differences, and the variance caused by genotype-by-environment interactions. See Falconer and Mackay 1996 and Lynch and Walsh 1998 for further details.

**xii: Many other attempts at integrating evolution and development have been made** (Huxley 1932, De Beer 1940, and Bonner 1958).

**xii: particularly in the past few decades** (Bonner 1982, Goodwin et al. 1983, Raff and Kaufman 1983, Arthur 1984, Nijhout 1991, Raff 1996, Gerhart and Kirschner 1997, Carroll et al. 2001, Davidson 2001, Wilkins 2002, Minelli 2003, West-Eberhard 2003, and Davidson 2006).

**xii: As William Bateson wrote in 1990** (Bateson 1900).

**xiii: Development, by contrast, is only weakly influenced by random events.** I am considering here only random events that perturb development, not variation in the environment that cues adaptive responses during development. A random event might be that the left side of a particular embryo experiences a slightly different temperature than the right side. In contrast, different embryos exposed to different temperatures may cue an adaptive response to temperature and develop to different sizes. This is called adaptive phenotypic plasticity (West Eberhard 1989, West-Eberhard 2003, DeWitt and Scheiner 2004).


## CHAPTER 1

**1: different levels of analysis provide complementary explanations for biological diversity** (Mayr 1961 and Sherman 1988).

**2: Different null mutations of the *Frigida* gene have arisen and spread at least twenty times and contribute to much of the variation in flowering time in natural populations of *Arabidopsis thaliana*** (Johanson et al. 2000,

Le Corre et al. 2002, Gazzani et al. 2003, Stinchcombe et al. 2004, and Shindo et al. 2005).

**3: Plant species closely related to *Arabidopsis thaliana* possess the *Frigida* gene** (Kuittinen et al. 2008).

**3: Null mutations appear to have spread in small populations that adapted rapidly to local conditions** (Toomajian et al. 2006).

**3: The null mutations reduce flowering time, but, since these mutations have not spread widely to other *Arabidopsis thaliana* populations, they are likely to confer disadvantages in other environments** (Korves et al. 2007 and Scarcelli et al. 2007).

**4: In a mutagenesis experiment to identify genes involved in controlling vernalization, only three out of about 50 mutations occurred in the *Frigida* gene** (Michaels and Amasino 1999; Scott Michaels, personal communication).

**4: mutations in at least 80 genes can affect flowering time** (Levy and Dean 1998).

## CHAPTER 2

**6: in Figure 2.3, we again see clear lines of descent.** Here I consider only the simple case of a gene lineage in the absence of recombination. Linkage and recombination add considerable fascinating complexity to population-genetic considerations. For further details, see Hill and Robertson 1966 and Barton 1995. For brief discussions of the effects of linkage on molecular evolution, see Gillespie 2006 and Nachman 2006.

**9: Figure 2.5.** Population genetic simulations were performed using PopG version 3.0, distributed for free by Joe Felsenstein at http://evolution.gs.washington.edu/popgen.

**12: in the human population, every possible mutation in the genome that is compatible with life has occurred about 240 times in the last generation** (Kruglyak and Nickerson 2001).

**14: Between human and mouse genes, for example, the ratio of nonsynonymous to synonymous substitutions is about 0.15. This implies that at least 85% of nonsynonymous mutations that arose on these two lineages were deleterious and were eliminated by purifying selection.** See Nei 2005 for data and discussion. Some nonsynonymous substitutions may have been positively selected. Positively selected mutations have a higher probability of fixation than neutral mutations in populations that are not too small (Kimura 1962). If some mutations were fixed by positive selection, then more than 85% of nonsynonymous mutations were deleterious and were eliminated by purifying selection.

**15: Figure 2.8.** Scanning electron micrograph of the wild type *Drosophila melanogaster* is by the author and of the *eyeless* mutant eye of *Drosophila melanogaster* was kindly provided by Peter Baatsen, staff scientist of the electron microscopy facility of the Center for Human Genetics, Leuven, Belgium. The photos of mouse embryos were kindly provided by Jack Favor and were adapted from Haubst 2006.

**16: Figure 2.9.** Scanning electron micrograph was created by Dr. Patrick Callaerts and is used with permission of *Science* magazine. © 1995 *American Association for the Advancement of Science.*

**19: purifying selection has acted on the DNA-binding domains of these transcription factors.** This argument predicts that the DNA binding domains of transcription factors will evolve more slowly than the target gene sequences bound by these proteins. This argument does *not* predict that transcription factor DNA binding domains will never evolve. In fact, in some cases, the DNA binding domains of transcription factors have evolved rapidly (Ting et al. 1998 and Jia et al. 2003, 2004). [In some cases, transcription factors as a class have evolved quickly (Bustamante et al. 2005), but it is not clear if this rapid evolution has occurred within or outside the DNA binding domains.] It also is likely that the rate of transcription factor evolution depends on the number of genes each transcription factor regulates and on the centrality of these processes to development (Hurst and Smith 1999).

## CHAPTER 3

**22: Geneticists define an allele as dominant when organisms containing one or two copies of the allele present the same altered phenotype.** This definition holds for genes on the autosomes. For genes on the X chromosome, dominance can be determined in the sex that carries two copies of the X chromosome (females in flies and humans). In the sex that carries one X chromosome (males in flies and humans), a single copy of a recessive allele can act alone, just like a dominant allele.

**24: Figure 3.2 shows an example of incomplete dominance for naturally occurring alleles in a population of beach mice.** This example comes from Steiner et al. 2007. These are mice carrying different genotypes for the *Agouti signaling protein* gene that were simultaneously homozygous for a "light" allele of the *Melanocortin 1 receptor* gene. Figure 5 from Steiner et al. 2007 shows that the "intermediate" phenotype for cheek pigmentation is apparent only on the homozygous "light" genotype of the *Melanocortin 1 receptor* gene. The *Agouti signaling protein* heterozygotes do not show an intermediate phenotype if mice simultaneously carry at least one "dark" allele of the *Melanocortin 1 receptor* gene. This phenomenon is called epistasis and is discussed in Chapter 5. In addition, patterns of dominance for these alleles vary, based on which trait is measured (Steiner et al. 2007). The photos in Figure 3.2 were kindly provided by Hopi Hoekstra.

**24: The majority of mutations that disrupt gene activity cause a quantitative reduction in the amount of gene product produced. These quantitative changes are unlikely to cause phenotypic dominance** (Wright 1929, Fisher 1930, and Kacser and Burns 1981).

**24: Even transcription factors can be considered enzymes; they act in pathways leading to the synthesis of specific RNAs** (Ptashne and Gann 1997 and Ptashne and Gann 2002).

**25: Figure 3.3.** The graphed functions come from equation (1) of Kacser and Burns 1981.

**25: Most enzymes are produced in sufficient quantity that their activity places them high on the plateau of flux through the pathway, toward the right of Figure 3.3.** Sewall Wright (1934) was the first to explore a potential physiological explanation for dominance. Kacser & Burns (1981) injected modern enzyme kinetics into Wright's modeling approach and my description of phenotypic dominance follows their paper.

**26: Figure 3.4.** Figure adapted from Figure 4 of Middleton and Kacser 1983. Data replotted with the standard errors that they report.

**27: Figure 3.5.** Scanning electron micrograph of *Antennapedia* mutation made by Rudolph Turner and used with his kind permission.

**27: in many organisms, the eye-lens crystallins . . . are enzymes, such as Lactate dehydrogenase protein and Argininosuccinase lyase protein** (Wistow et al. 1987, Piatigorsky et al. 1988, and Wistow 1993).

**27: high levels of expression resulted from mutations that increased expression specifically in the lens, without any changes that altered the catalytic activity of the enzymes** (Piatigorsky and Wistow 1989, Piatigorsky and Wistow 1991, and Yang and Cvekl 2005).

**27–28: new genes arise frequently by mutations that duplicate existing genes.** Gene duplications can arise through several different mechanisms. See Chapter 10 of Li 1997 and Zhang 2003, for further details. In addition, some gene duplication events cause the fusion of two different genes, which generates chimeras, potentially creating a new functional gene in one step (Long and Langley 1993, Wang et al. 2002, Jones and Begun 2005, Suetsugu et al. 2005, and Rogers et al. 2009).

**28: Null mutations . . . are often effectively neutral, since the second copy of the gene is still present.** See Kondrashov and Koonin 2004 for a discussion of a possible connection between genetic dominance and gene duplication.

**28: Gene duplication, therefore, initiates a lot of evolutionary novelty at the molecular level** (Ohno 1970, Holland 1999, Zhang 2003, Jaillon et al. 2004, and Kellis et al. 2004).

**28: in many cases eye-lens crystallins have evolved from enzymes after gene duplication** (Wistow 1993). In addition, in some cases it appears that genes first evolved to encode both enzymes and lens crystallins and then duplicated and divided these roles (Piatigorsky and Wistow 1991 and Wistow 1993).

**28: some fish that live in subzero waters have evolved antifreeze proteins that capture ice in the bloodstream to prevent ice crystals from puncturing their cells** (Chen et al. 1997).

**30: Many transcription factors can act as repressors or as activators depending on physical interactions with other proteins.** Activation or repression by transcription factors usually involves additional proteins that I do not discuss here. See Ptashne and Gann 2002 for an introduction to this problem.

**31: These weak binding interactions between transcription factors can greatly increase the binding affinity of transcription factors for DNA.** See Ptashne 2004 for an introduction to this problem.

**31: These stripes of expression help to establish the segmented body of the fly.** See Lawrence 1992 and Carroll et al. 2001 for an introduction to *Drosophila melanogaster* embryonic development.

**31: The *cis*-regulatory region driving expression of the second most anterior stripe (stripe 2) has been studied in considerable detail** (Small et al. 1991 and Small et al. 1992).

**34: relative fitness—the reproductive success of an individual in comparison with other individuals in the population.** Fitness is difficult to measure in most organisms, for the obvious reason that it is difficult to count the total number of surviving offspring in natural conditions. For long-lived organisms, measuring fitness requires tracking reproductive success over many years. In contrast, short-lived organisms are often small and often produce large numbers of offspring. Most succumb prior to reproduction. It can be very challenging, to say the least, to identify the remaining surviving offspring. Instead, investigators usually measure phenotypic attributes that are thought to be correlated with fitness, such as fecundity or body size. See Endler 1986 for a thorough discussion of different definitions of fitness and a review, up to 1984, of estimates of natural selection in the wild. See Kingsolver et al. 2001 for a more recent review of studies of natural selection in the wild.

**35: semidominance, where the heterozygote enjoys a fitness exactly intermediate between the two homozygotes: $h = 0.5$.** In this case, the effects of the alleles are additive. This also is called genic selection (Hartl and Clark 1997).

**35: I have not illustrated several other important cases, namely when the heterozygote experiences higher or lower fitness than both of the homozygotes.** When the heterozygote enjoys higher fitness than both homozygotes—a situation called heterozygote advantage—both alleles can be preserved in the population by selection. Examples include the maintenance of a haemoglobin polymorphism that causes sickle-cell anemia in human populations (Allison 1954) and the maintenance of plumage color polymorphism in buzzards (Kruger et al. 2001).

**37: Figure 3.9.** Pictures kindly provided by Rowan Barrett.

**37: variation at one gene, called *Ectodysplasin*, has a strong effect on the number of bony plates** (Peichel et al. 2001 and Colosimo et al. 2005).

**38: Figure 3.10.** Adapted from Figure 2 of Barrett et al. 2008.

**38: phenotypic dominance displayed a poor correlation with fitness dominance.** These results come from Barrett et al. 2008. As these authors note, these results are unlikely to be caused by differential effects of changes in the number of armor plates at different life stages. Instead, it seems more likely that they result either from pleiotropic effects of *Ectodysplasin* on other phenotypic traits or from effects of one or more loci closely linked to *Ectodysplasin*.

**39: They reduce fitness only when organisms are homozygous for the allele, and homozygotes are produced only when heterozygotes mate, which happens infrequently when alleles are rare.** Inbreeding will increase the frequency of homozygotes, revealing these deleterious recessive effects (Scott et al. 1995).

**40: Figure 3.12.** The curves for this figure are drawn from equation 6.2.6 of Crow and Kimura 1970 assuming a constant mutation rate ($10^{-6}$).

**40: all populations harbor many deleterious mutations at low frequency** (Crow and Simmons 1983).

**40: mutation–selection balance.** Mutation–selection balance will be achieved at equilibrium, which assumes that the population size remains constant. The equilibrium frequency depends on multiple parameters, including dominance, population size, population structure, and genetic linkage (Fisher 1930, Haldane 1937, Fisher 1958, Lande 1975, and Hastings 1989). In addition, the sex chromosomes will reach a different equilibrium frequency than the autosomes. For a general introduction to these issues see pp. 258–262 of Crow and Kimura 1970 and pp. 236–240 of Hartl and Clark 1997.

**42: The probability that a selectively advantageous semidominant mutation is ultimately fixed is equal to approximately twice the selection coefficient enjoyed by the heterozygote (approximately 2*hs*).** This is an approximation that assumes that the population size is reasonably large and that the selection coefficient is relatively small (Haldane 1927, Kimura 1962, and Gillespie 2004). As long as the population size is large enough that changes in gene frequency can be approximated as a continuous stochastic process, the probability of fixation for a new semidominant allele, where the heterozygote has a fitness of 1 + 0.5*s*, can be estimated as

$$\frac{1-e^{-\frac{N_e s}{N}}}{1-e^{-2N_e s}},$$

where $N$ is the population size and $N_e$ is the effective population size. See Crow and Kimura 1970, section 8.8.3, for further details. This result was originally derived for a new semidominant mutation (Kimura 1962). Crow and Kimura 1970 give a more exact result for an allele with arbitrary $s$ and $h$ as their equation 8.8.3.21. For a completely recessive allele ($h = 0$), and assuming small and positive $s$, large $N_e s$, and $N_e = N$, the probability of fixation is approximately

$$\sqrt{\frac{2s}{\pi N}}$$

(equation 9.8.3.24 of Crow and Kimura 1970).

**42: In contrast, a new neutral allele has a probability of**

$$\frac{1}{2N}$$

**of ultimately being fixed.** A new neutral allele confers the same fitness, on average, as the original allele, so each copy of a gene has the same probability of completely replacing all other copies. For a new mutation, its current frequency is

$$\frac{1}{2N},$$

where $N$ is the number of individuals, so the probability that this allele completely replaces all others is

$$\frac{1}{2N}.$$

**43: The stronger the dominance of an allele, the more likely that it will be fixed by natural selection.** The probability of fixation is approximately directly proportional to the degree of dominance and is approximately equal to $2hs$, where $h$ is the degree of dominance and $s$ is the homozygote fitness advantage. See section 3.9 of Gillespie 2004 and Orr and Betancourt 2001.

**43: The effect of dominance on the probability of fixation of new mutations was discovered by J. B. S. Haldane and is called "Haldane's sieve" in his honor** (Haldane 1924, 1927). Apparently, this finding was given the name "Haldane's sieve" by J. R. G. Turner (Turner 1981).

**44: Figure 3.13.** Simulations were performed using PopG software.

**45: Figure 3.14.** Photograph of melanic and typical forms of *Biston betularia* © J. L. Mason/ardea.com and used with permission.

**45: peppered moths, *Biston betularia*, come in multiple morphs . . . The alleles conferring the dark morph are dominant to the alleles conferring**

**the light morph.** There are multiple color morphs in nature: the light form; multiple intermediate forms; called *Biston betularia insularia*; and the darkest form, called *Biston betularia carbonaria* (Cook 2003). Only the light form and *Biston betularia carbonaria* are shown in Figure 3.14. All morphs appear to be alleles of the same gene (Lees 1968). As Grant 2004 showed, alleles causing melanism of British populations (*Biston betularia carbonaria*) occur at the same gene as alleles causing melanism in American populations (*Biston betularia swettaria*).

**46: both dark morphs suffered a similar fitness decrement** (Cook et al. 1986).

**46: At first sight, it might seem that Haldane's sieve predicts that adaptation should occur most frequently by fixation of dominant alleles.** In fact, this prediction has been made several times (Turner 1981, Charlesworth 1992, and Noor 1999).

**46: In a variable environment, neutral or deleterious alleles that were already present in a population may suddenly become advantageous.** Kimura 1983 termed the transition of a previously neutral allele to an advantageous allele the Dykhuizen–Hartl effect, after their 1980 paper (Dykhuizen and Hartl 1980), which showed that previously neutral mutations can become advantageous when either the genetic background or the environment changes.

**47: Under a wide range of assumptions, the probability of fixation for preexisting alleles maintained by mutation–selection balance is independent of dominance.** This result was shown by Orr and Betancourt 2001, and it holds when the degree of dominance is not close to 0. Under mutation–selection balance, alleles derived from independent mutations but with similar phenotypic effects may be present in the population. In this case, selection may sweep several of these alleles to high frequency. "Fixation," in this case, involves elimination of the ancestral allele, but "substitution" with a set of functionally equivalent alleles (Orr and Betancourt 2001, Innan and Kim 2004, Hermisson and Pennings 2005, Przeworski et al. 2005, and Pennings and Hermisson 2006). This is sometimes called a "soft sweep" to contrast it with the "hard selective sweep" of a single new mutation. Given enough time, one of these alleles is expected to become fixed by genetic drift. Some possible empirical examples of soft sweeps include the apparent fixation of multiple alleles of the *tan* locus in *Drosophila santomea* (Jeong et al. 2008) and the selection of what appear to be two alleles of the *Ectodysplasin* locus in independent populations of sticklebacks (Colosimo et al. 2005).

**47: insects carrying mutations conferring insecticide resistance suffer reduced fitness in the absence of the insecticide.** Australian sheep blow flies (McKenzie 1990) and mosquitoes (Rowland 1988) carrying mutations in a subunit of the GABA receptor that confer resistance to dieldrin have reduced fitness in the absence of dieldrin. Some insecticide-resistance alleles in the mosquito *Culex pipiens* also cause reduced fitness (Chevillon et al. 1997). See review by Andreev et al. 1999.

**47: adaptation in the long term will make increasing use of alleles intro-
duced by mutation.** Orr and Betancourt 2001 discuss this scenario and the difficulties associated with testing related predictions. Barton and Keightley 2002
show that, while a changed environment initially selects on the standing variation,
new mutations begin to contribute to the response to natural selection surprisingly quickly.

CHAPTER 4

**49: a mutation that inactivates the *white* gene, causes the fly's normally
red eyes to be white, but also removes pigmentation from the male
testis, changes the shape of the female spermatheca, and reduces lifespan and viability** (Dobzhansky 1927 and Dobzhansky and Holz 1943).

**50: In 1927, Theodosius Dobzhansky considered these to be pleiotropic
effects of one gene** (Dobzhansky 1927).

**50: this gene encodes a starch-branching enzyme and that the wrinkled
allele is null, producing no active enzyme** (Bhattacharyya et al. 1990).

**51: In 1922, Hermann Muller noted that different mutations of the *trun-
cate* gene in *Drosophila melanogaster* caused shortened wings, eruption of
the thorax, and lethality, or some combination of these phenotypic
effects** (Muller 1922).

**51: The pleiotropic roles of genes can result either from a single protein
performing multiple distinct molecular jobs or from expression of a
protein in multiple tissues.** The word "protein" can be replaced with "gene
product" in this and the next paragraph and these paragraphs remain true. Some
genes encode functional RNA molecules rather than proteins.

**51–52: the *Ultrabithorax* gene . . . is required for normal development of
many organs of the thorax and abdomen,** (Lewis 1978 and Casanova et al.
1985) **including: the balancer organs called halteres on the third thoracic
segment** (Weatherbee et al. 1998 and Roch and Akam 2000); **the second and
third pairs of legs** (Stern 2003 and Davis et al. 2007); **the shape and size of
the third thoracic and first abdominal segment** (Kerridge and Morata 1982,
Casanova et al. 1985); **and organs of the nervous system and the gut** (Bienz
et al. 1988, Immergluck et al. 1990, and Rozowski and Akam 2002).

**52: Figure 4.1.** Adapted from Tour et al. 2005.

**52: The Ultrabithorax protein represses transcription of some genes in
some tissues** (Galant and Carroll 2002, Galant et al. 2002, Gebelein et al. 2002,
and Hersh and Carroll 2005); **and activates transcription of some genes in
others** (Wiellette and McGinnis 1999 and Tour et al. 2005).

**52: These different repressive and activator functions involve different
domains of the protein** (Tour et al. 2005).

**52: deletion of 24 amino acids from the end of the Ultrabithorax protein causes subtle phenotypic effects** (Hittinger et al. 2005).

**53: Figure 4.2.** Adapted from Galant and Carroll 2002, Ronshaugen et al. 2002, and Hittinger et al. 2005.

**54: Figure 4.3.** Photo of wild-type fly kindly provided by Benjamin Prud-homme and Nicolas Gompel and photo of *bithorax* mutant fly by Ed Lewis.

**54: all of these domains are highly conserved across insects** (Tour et al. 2005).

**55: Figure 4.4.** Adapted from results reported in Geyer and Corces 1987 and Geyer and Corces 1992.

**55: Expression of Yellow protein in these spots is not sufficient to generate pigmentation.** Wittkopp et al. 2002 showed that ectopic expression of Yellow gene product is not sufficient to generate pigmentation in all body regions. Gompel et al. 2005 report that ectopic expression of Yellow gene product in the *Drosophila melanogaster* wing is not sufficient to generate pigmentation in the wing.

**56: Figure 4.5.** Adapted from Gompel et al. 2005 and photographs courtesy of Nicolas Gompel and Benjamin Prud'homme.

**57: Figure 4.6.** Adapted from Gompel et al. 2005 and fly wing images courtesy of Nicolas Gompel and Benjamin Prud'homme.

**58: Figure 4.7.** Adapted from Prud'homme et al. 2006 and fly wing images courtesy of Nicolas Gompel and Benjamin Prud'homme.

**59: At least two and no more than seven point mutations generate this difference between the species** (Prud'homme et al. 2006).

**60: If a gene has pleiotropic roles, then we need to consider the possibility that mutations in this gene will have pleiotropic effects.** Mutations in a nonpleiotropic gene may confer an additional novel role on a gene, thus conferring pleiotropic roles upon the gene.

**60: Many genes are known to contribute to development of multiple organs, and mutations in these genes may have pleiotropic effects**. For example, see the discussion of the *decapentaplegic* gene in Stern 2000.

**62: Figure 4.9.** Drawings of flies were reproduced by permission of CSIRO Australia © CSIRO from *The Insects of Australia – Volume 2.*

**62: But many mutations, such as many mutations that alter protein function, will simultaneously affect many traits.** Of course, some *cis*-regulatory mutations may have pleiotropic effects (Argeson et al. 1996 and Clark et al. 2006) and some protein-coding mutations may have specific effects (Lynch and Wagner 2008).

**62: Pleiotropic mutations may move individuals in a random direction through this three-dimensional space, but most steps in a random direc-**

**tion will tend to move the phenotype further away from the optimum.**
This geometric model of adaptation was introduced originally by R. A. Fisher
(1930). Fisher modeled adaptation as a mutational step from the mean phenotype
of a population toward a phenotype that confers higher fitness. Fisher assumed
that the population sits on the surface of a hyperdimensional sphere—a sphere
defining phenotypes of equal fitness—where the optimal phenotype resides at the
center of the sphere. High dimensionality results from the pleiotropy of mutational effects. Fisher's main point was that, in a high-dimensional space, mutations
of large effect have a vanishingly small probability of improving fitness. Mutations
of small effect, however, approximate the case of a single dimension, where mutations in one direction improve fitness and those in the other direction decrease fitness. Fisher showed that the probability of improvement falls off quickly with
increasing magnitude of the mutational effect. Fisher pointed out that the probability that a mutation increases fitness is determined by the size of the mutational
effect relative to what he called the standard magnitude of change,

$$\frac{d}{\sqrt{n}},$$

where $d$ is the diameter of the sphere and $n$ is the dimensionality of the sphere. As
the dimensionality increases, the probability that a pleiotropic mutation of any
given size will be favorable diminishes. (Fisher considered high dimensionality to
reflect more complex adaptations, which conflates mutational effects and gene
roles. See Stern 2000 for further discussion.)

**63: This assumption has been partially tested in the yeast** *Saccharomyces
cerevisae* (Cooper et al. 2007).

**63: Four thousand seven hundred and eighteen strains of yeast were
generated, each of which carried a deletion of a different single gene.**
These deletions completely eliminated gene function. While many mutations can
eliminate gene function, it is not yet known whether mutations that do not
destroy gene function also show a negative relationship between pleiotropy and
fitness.

**64: mutations with fewer pleiotropic effects are more likely to improve
fitness than are mutations having extensive pleiotropic effects.** See also
theoretical models of Baatz and Wagner 1997, Orr 2000, Griswold and Whitlock
2003, Welch and Waxman 2003, and Otto 2004, and further discussion by Stern
2000 and Stern and Orgogozo 2008.

**64: So far, we have considered only the probability that a single muta-
tion will be beneficial.** Kimura (1983) recognized that Fisher's geometric
model of adaptation quantified only the probability that a mutation is beneficial
(Figure A, below). Kimura pointed out that this probability must be multiplied by
the probability of substitution of a mutation of effect size $s$ to find the size distribution of mutations fixed during adaptation. The probability of substitution is
approximately *2hs*. Thus, while Fisher found that mutations of very small effect

tend to be favorable, these mutations have a low probability of substituting. Kimura claimed that mutations of intermediate size are therefore expected to fix more frequently during adaptation (Figure B, below).

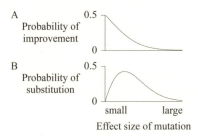

It is a little difficult to translate these estimates of "small," "intermediate," and "large" into practical terms. In these models, the effect size of a mutation is a standardized measure:

$$r\frac{\sqrt{n}}{d},$$

where $r$ is the mutational step size, $n$ is the dimensionality of the phenotypic trait resulting from mutations of pleiotropic effect, and $d$ is the diameter of the $n$-dimensional hypersphere, on the surface of which the population currently sits. Thus, a mutation of a given size may be considered "small" in a low-dimensional space and "large" in a high-dimensional space. This seems to jibe with our intuition that mutations of a given phenotypic effect size are less likely to be adaptive if they have pleiotropic effects. See Orr 1998, Barton and Keightley 2002, and Waxman and Welch 2005 for additional discussion.

**65: We might visualize adaptation as a walk toward the fitness optimum, with multiple mutations substituting in the population over time.** Orr (1998) recognized that Kimura's representation of adaptation was incomplete, since adaptation probably usually requires substitution of multiple mutations (Barton and Keightley 2002). He therefore calculated the distribution of mutational-effect sizes expected during an adaptive walk (in terms of the standardized measure of mutation size,

$$r\frac{\sqrt{n}}{d}).$$

**65: many studies have demonstrated that the individual phenotypic differences between species are usually caused by multiple mutations, usually found in multiple genes** (Orr 2001).

**65–66: As a population evolves closer to the optimum, mutations of smaller and smaller effect, on average, will tend to be substituted, since the distance to the optimum gets smaller with each successive substitution event.** Orr (1998) showed that, under Fisher's geometric model, the distribution of effect sizes of mutations substituted is expected to follow an exponential distribution. That is, we expect to observe a few mutations with medium to large effects and a large number of mutations with small effects. See also Barton and Keightley 2002, Waxman and Welch 2005, and Kopp and Hermisson 2009.

**66: In the nematode worm *Caenorhabditis elegans*, the *lin-48* gene . . . is expressed in the hindgut, the excretory duct, the male tail, and in a variety of other cells** (Johnson et al. 2001).

**66: Loss of *lin-48* protein expression in the excretory duct of *Caenorhabditis briggsae* was caused by at least four mutations in a *cis*-regulatory region of the *lin-48* gene** (Wang and Chamberlin 2002).

**67: Figure 4.13.** Adapted from Wang and Chamberlin 2002.

**69: This morphological innovation in *Drosophila sechellia* was caused by evolutionary changes at the *shavenbaby* gene** (Sucena and Stern 2000).

**69: the complex expression pattern of the Shavenbaby gene product is determined by at least three enhancers** (McGregor et al. 2007).

**70: Figure 4.15.** Adapted from McGregor et al. 2007.

**71: This generates an asymmetry; multiple inputs regulate transcription of the *lin-48* and *shavenbaby* genes, and their gene products then regulate transcription of multiple other genes** (Wray et al. 2003).

**72: Gene duplication is the wellspring of most new genes** (Muller 1935 and Ohno 1970). New genes also can arise from the fusion of two genes, forming chimaeric genes (Wang et al. 2002, Jiang et al. 2004, Suetsugu et al. 2005, and Rogers et al. 2009). New genes also arise *de novo*, from nonfunctional DNA (Cai et al. 2008, and Knowles and McLysaght 2009).

**72: most duplicated genes go extinct shortly after they first appear** (Lynch and Conery 2000, Zhang 2003, and Rogers et al. 2009).

**73: this model is unlikely to explain the abundance of duplicated genes in nature.** See Force et al. 1999 for a more detailed discussion of difficulties with models that rely entirely upon new adaptive mutations for the fixation of gene duplicates.

**73: This process is named the duplication–degeneration–complementation model** (Force et al. 1999 and Lynch and Force 2000).

**74: The fixation of a neutral allele takes an average of $4N$ generations** (Kimura and Ohta 1969), where $N$ is the effective (not the demographic) population size, which is discussed further in Chapter 6.

**75: Figure 4.17.** Adapted from Stauber et al. 1999.

**75: The model therefore works best when degenerated neutral alleles drift through populations quickly.** For two reasons, duplicated genes are more likely to survive in smaller populations. First, duplicates that are slightly deleterious in a large population may be effectively neutral in smaller populations (see Chapter 6). Thus, the duplicate will not be eliminated quickly by selection. Second, smaller populations are more likely to retain duplicates through the duplication–degeneration–complementation process.

**76: Figure. 4.18.** Adapted from Stauber et al. 2002.

**77: One striking feature of *bicoid* gene evolution is that evolution of novel functions required changes to amino acids that are conserved in many related genes.** The "related" genes referred to here are *Hox* genes (Stauber et al. 1999).

## CHAPTER 5

**79: In traditional genetic terms, epistasis is said to occur when the effects of one allele hide the effects of an allele at another locus.** This definition comes from William Bateson (1909).

**80: Epistasis tests provide a powerful method for detecting interactions between gene products during development.** Epistasis tests provide a reasonably reliable method for determining the relative position of a gene in a genetic network, but only if the test is performed with null alleles or constitutively active alleles (Avery and Wasserman 1992). Alleles of intermediate state—for example, those with partial loss of function—can produce misleading results in an epistasis test.

**82: In population genetics, modification of allelic effects by variation at other loci is also called epistasis, and it can be detected when the effect of combining alleles at different loci in a single individual does not equal the sum of the effects of each allele alone.** This definition of what he called epistacy comes from R. A. Fisher (1918).

**83: Figure 5.4.** Adapted from Figure 5 of Steiner et al. 2007.

**83: Epistasis in the population genetics sense has been detected in a wide variety of organisms in nature.** Some recent examples include Kroymann and Mitchell-Olds 2005, Gerke et al. 2006, Steiner et al. 2007, and Gerke et al. 2009. For a book-length review, see Wolf et al. 2000.

**84: mutations may substitute in a population and display epistasis with respect to mutations that substituted in the past.** In this book, I do not discuss the contribution of serial epistasis to postzygotic reproductive isolation. When sexual populations remain isolated for a sufficiently long period of time, they accumulate substitutions, either through natural selection or by genetic drift. If individuals from these isolated populations attempt to reproduce, genes that have accumulated those mutations may fail to work properly in the mixed genetic background and may cause reproductive incompatibilities or worse. This is called the Bateson–Dobzhansky–Muller model for the evolution of reproductive incompatibilities (Orr 1996 and Coyne and Orr 2004).

**84: at a population–genetic level, developmental constraint appears to reflect simply a long history of serial epistasis.** The literature on developmental constraints is, to my mind, both a quagmire and a minefield. Various authors parse the term in ever so subtly different ways, but few have defined the term in ways that are easily accessible by population genetics. The field is a minefield because many people have surprisingly strong feelings about the presence, or absence, of developmental constraints both in general and in specific cases. The clearest definition of developmental constraints comes from a paper coauthored by a group of developmental and evolutionary biologists (Maynard Smith et al. 1985). They defined developmental constraints as "biases on the production of variant phenotypes or limitations on phenotypic variability caused by the structure, character, composition, or dynamics of the developmental system." I think most, and possibly all, of the examples of sources of developmental constraints discussed in this paper reflect serial epistasis. In addition to developmental constraints, populations may fail to evolve in a particular direction simply because the relevant genetic variation has not yet appeared, even if the evolutionary transition requires only a single mutation.

**85: New adaptive mutations tend to sweep through populations relatively quickly, so fitness improvements usually result from the serial fixation of new mutations.** But not always. If multiple mutations arise on the same nonrecombining chromosome before fixation, then the fitness effects of the different mutations may interfere with each other. This is called clonal interference (Kim and Orr 2005, de Visser and Rozen 2006, and Kao and Sherlock 2008) and it was first noted by both Fisher (1930) and Muller (1932).

**85: Sometimes, replicate populations evolve to different maximal fitness levels, implying that different populations have followed different mutational trajectories to improved fitness and that some paths lead to higher fitness than others.** Reviewed in Elena and Lenski 2003.

**86: Blowflies resistant to diazinon were first detected in 1965, and the resistance allele spread quickly through most of the population.** Much of this work is reviewed in Batterham et al. 1996. See McKenzie and Clarke 1988 and Davies et al. 1996 for further details.

**86: Approximately 470 million years ago, a gene duplication in the vertebrate lineage generated two genes that evolved into the *Glucocorticoid Receptor* gene and the *Mineralocorticoid Receptor* gene.** The ancient gene duplication is evident on the phylogeny shown in Figure 5.5 because the two human genes (the *Glucocorticoid Receptor* gene and the *Mineralocorticoid Receptor* gene) appear on divergent lineages, and the human *Glucocorticoid Receptor* gene lineage is intersected by lineages leading to *Glucocorticoid Receptor* genes from divergent vertebrates. Figure 5.5 adapted from Ortlund et al. 2007.

**86–88: The Glucocorticoid Receptor protein binds the adrenal steroid cortisol and is required in humans for glucose homeostasis and for proper regulation of the stress response, inflammation, and immunity, as**

well as for other functions. The Mineralocorticoid Receptor protein, in contrast, is strongly activated by aldosterone, and it regulates electrolyte homeostasis, kidney and colon function, as well as other functions (Bentley 1998).

**88: Studies of the activation patterns of these resurrected ancestral proteins . . . allowed identification of the individual amino-acid substitutions that contributed to the specificity of the Mineralocorticoid Receptor protein for aldosterone and of the Glucocorticoid Receptor protein for cortisol** (Ortlund et al. 2007).

**90: Figure 5.6.** Adapted from Ortlund et al. 2007. Joe Thornton kindly redrew this figure for this book.

**91: A known mutation is then introduced into each strain, usually by introgression.** Introgression involves the transfer of genetic material from one strain into another by hybridization and repeated backcrossing to one of the parental strains. At each generation, a marker—such as a molecular marker, or a phenotypic marker, or, as in these experiments, the phenotypic effect of a mutation of interest—is tracked to ensure that the DNA region of interest is perpetuated through each generation of backcrossing.

**92: Figure 5.7.** Scanning electron micrographs of wild type and *Antennapedia* heads by Rudi Turner. Remainder of figure adapted from Gibson et al. 1999.

**93: Figure 5.8.** Images by the author.

**93: Figure 5.9.** Adapted from Gibson and Helden 1997.

**93: As the mutation traverses the population, it may encounter its own unique spectrum of epistatic interactions.** A special case occurs when an advantageous mutation arises that is closely linked to one or more deleterious mutations. In such a case, the fitness effects of the advantageous and deleterious mutations interfere with one another. This is called the Hill–Robertson effect (Hill and Robertson 1966), and it is thought to contribute to the prevalence of sexual reproduction and recombination, since recombination reduces linkage disequilibrium (Felsenstein 1974).

**94: any combination of phenotypic characteristics could interact to cause fitness epistasis.** See Lunzer et al. 2005 for an example, at the molecular level, of how additive phenotypic effects can result in epistatic fitness effects.

**94: Here is an example from garter snakes** (Brodie 1992).

**94: Many mutations for pigmentation patterns are unlikely to influence escape behavior.** There are at least two examples known of mutations that influence both pigmentation and behavior. In *Drosophila melanogaster*, null mutations in the *yellow* gene reduce pigmentation (Wittkopp et al. 2002). Flies carrying mutations in the *yellow* gene also perform courtship at lower rates than flies carrying a functional allele of the *yellow* gene (Sturtevant 1915 and Bastock 1956). This behavior apparently requires expression of the Yellow protein in four cells in the larval central nervous system (Drapeau et al. 2006). Other *Drosophila* species also require a functional *yellow* gene for both pigmentation and courtship behavior (Rendel 1945 and da Silva et al. 2005). In a second example, in mice and humans, mutations in the *Melanocortin 1 Receptor* gene alter pigmentation (Rees 2003) and modulate response to pain (Mogil et al. 2003). Despite these pleiotropic roles of the *yellow* gene and the *Melanocortin 1 Receptor* gene, many mutations in either gene can have specific effects either on pigmentation or on behavior. For example, the behavioral role of the *yellow* gene is controlled by a 300 bp *cis*-regulatory sequence, and mutations in this region might affect only behavior (Drapeau et al. 2006). Mutations in the other *cis*-regulatory regions shown in Figure 4.4 can have specific effects on pigmentation without disrupting behavior (Drapeau et al. 2006).

**95: Figure 5.10.** Adapted from Brodie 1992. Figure kindly provided by Edmund Brodie.

**97: Epistasis of standing variation is widespread in all natural populations, and it contributes to every imaginable phenotype.** For reviews, see Whitlock et al. 1995, Wolf et al. 2000, and Malmberg and Mauricio 2005. Some recent examples include Kroymann and Mitchell-Olds 2005, Tatsuta and Takano-Shimizu 2006, Korves et al. 2007, Steiner et al. 2007, and Gerke et al. 2009.

**97: The prevalence of epistasis suggests that it has the potential to play some role in evolution. The question is, what role or roles?** For thorough and varied discussions of the possible roles of epistasis in evolution see Whitlock et al. 1995, Wolf et al. 2000, Moore and Williams 2005, and Weinreich et al. 2005.

**97: epistasis of standing variation plays an insignificant role in adaptation, since natural selection cannot act efficiently on epistatic variation.** A simple way to see this is to recognize that an adaptive allele with epistatic effects may increase fitness in only some genetic backgrounds. The net selection for this allele will therefore be less than for an allele with additive effects, which increases fitness, on average, in all genetic backgrounds. Quantitative genetics provides another way to think about this problem (Falconer and Mackay 1996 and Lynch and Walsh 1998).

**97: The obvious potential for epistasis to generate novel variants, however, has attracted adherents repeatedly—most famously, Sewall Wright—to the proposal that epistasis of standing variation provides a key source of phenotypic novelty in natural populations.** Sewall Wright was the first to propose this idea in what he called the Shifting Balance Theory (Wright 1931, 1932, 1980). See Coyne et al. 2000 for a critique of the Shifting Balance Theory. Other views on the importance of epistasis to evolution are discussed thoroughly in Wolf et al. 2000.

**97: A mutation with pleiotropic effects may alter multiple phenotypic features during the same stage of life; or, the pleiotropic effects may manifest themselves at different life stages** (Scarcelli et al. 2007).

**97: A pleiotropic mutation may even have different phenotypic effects when the organism experiences different environments** (Wilczek et al. 2009).

**98: By increasing the variation in fitness associated with a particular mutation, epistasis effectively reduces the selection coefficient associated with a mutation.** Increased variance in reproductive success effectively causes a reduction in the effective population size, $N_e$. The probability of fixation of a new advantageous mutation depends, in a complicated way, both on $N_e$ and on $N$: for example, the probability of fixation of a new semidominant mutation is approximately

$$\frac{1-e^{-\frac{N_e s}{N}}}{1-e^{-2N_e s}}.$$

If $N$ and $N_e$ are equal, then the probability of fixation is relatively insensitive to population size for large populations, but the probability of fixation increases in very small populations (approximately $N < 100$). But when $N_e$ and $N$ are unequal, the probability of fixation is approximately $2hsN_e/N$ (see equation 8.8.3.19 of Crow and Kimura 1970 and section 3.9 of Gillespie 2004). Thus, mutations that increase the variance in reproductive success—by any mechanism, including epistasis—may have a lower probability of substituting than mutations that do not alter the variance in reproductive success (Crow and Kimura 1970; Gillespie 2004).

**98: There are many examples of presumptive serial epistasis between the mutations that fixed during evolution.** It is usually assumed that mutations causing phenotypic differentiation of species were substituted serially. This assumption seems more reasonable than the alternative—that the mutations segregated simultaneously prior to substitution of both—for the simple reason that substitutions tend to occur quickly relative to the total divergence time between closely related species. See Jeong et al. 2008 for a possible example of simultaneous substitution of multiple alleles of the *tan* locus. Examples of presumptive serial epistasis include loci for *Drosophila* cuticular hydrocarbon profiles (Noor and Coyne 1996, and Coyne and Charlesworth 1997), *Drosophila* sperm and testis length (Joly et al. 1997), *Nasonia* wing size (Gadau et al. 2002), *Drosophila* pigmentation (Carbone et al. 2005), *Drosophila* ovariole number (Orgogozo et al. 2006), and *Drosophila* male sex comb tooth number (Tatsuta and Takano-Shimizu 2006). Perhaps more surprisingly, sometimes serial epistasis is not detected or is weak for loci that differentiate species. Examples include quantitative traits in tomato (deVicente and Tanksley 1993), *Drosophila* pigmentation (Ng et al. 2008), and *Drosophila* mating preferences (Moehring et al. 2006).

CHAPTER 6

**101: predators usually have smaller populations than their prey.** Prey always have greater productivity than predators, leading to the familiar ecological pyramids (Odum et al. 2005). This usually translates into predator populations being smaller than prey populations. But, in some cases, predator populations can be larger. For example, herbivorous insect populations are sometimes much larger than the populations of their plant prey.

**102: large populations have a wider spectrum of mutations available for natural selection to act upon.** See the Discussion section of Liu et al. 1996 for a thoughtful discussion of this and related topics.

**102: When we wish to evaluate the role of population size in evolution, it is not enough simply to consider the census population size.** For many purposes in population genetics, the effective population size is more useful than is the census population size. However, in some cases, such as when estimating the probability of fixation, estimates of both the effective and the census population sizes are required.

**102: genetic drift is more important in small populations than in large ones.** This result of theoretical investigations has been tested several times. Weber 1990 and Weber and Diggins 1990 provide reasonably compelling evidence that increasing effective population size—at least up to an effective size of approximately 480—leads to an increased response to selection in the laboratory. One possible complication is that these selection experiments were carried out for more than 50 generations and new mutations are thought to contribute an increasing amount of the available genetic variation after about generation 20 (Barton and Keightley 2002). Large populations will accumulate more genetic variation through new mutations than will small populations and this may have contributed to the response to selection observed in these experiments (Weber and Diggins 1990).

**103: Male sage grouse . . . gather in large groups called leks to display to females in the hopes of securing one or more mates.** Sage grouse are a famous lekking species, but many species of birds, mammals, and insects form leks (Wiley 1991 and Högland and Alatalo 1995).

**103: Males show a large variance in reproductive success** (Högland and Alatalo 1995 and Mackenzie et al. 1995).

**103: We can *estimate* the effective population size . . . by examining its effect on genetic drift.** For further discussion, see section 7.6 of Crow and Kimura 1970, chapter 7 of Hartl and Clark 1997, or section 2.7 of Gillespie 2004. See Gillespie 2006 for a brief, but clear, explanation of stochastic processes in evolution.

**103: The effective population size is the size of a hypothesized ideal population . . . that would experience the same amount of genetic drift as observed in our real population.** This is sometimes called the variance effective population size. Inbreeding causes an increase in the frequency of homozygotes, and this effect is captured in a quantity called the inbreeding effective population size. See Chapter 7 of Crow and Kimura 1970 for further discussion of effective population size estimates.

**103: Estimates of the effective population size of natural populations are typically smaller, and often much smaller, than the census size** (Frankham 1995).

**103–104: The reduced genetic drift in large populations relative to small populations allows selection to be more discriminating in large populations.** Selection for a mutation tends to dominate over genetic drift when the product of twice the effective population size times the selection coefficient associated with a mutation ($2N_e s$) is either greater than 1 or less than $-1$. When $2N_e s$ is between $-1$ and 1, drift tends to predominate. Positive selection prevails when $2N_e s$ is greater than 1 and negative (purifying) selection acts when $2N_e s$ is less than $-1$. See section 3.9 of Gillespie 2004.

**104: There are therefore two possible reasons why a mutation may be effectively neutral . . . : either the selection coefficient is too small, or the population size is too small.** The rate of adaptive evolution is the product of the rate at which new adaptive mutations are introduced, which is the product of the population size and the mutation rate ($Nu$), and the probability of their substitution, which also depends on population size (see note on pp. 190–191 on the probability of substitution). See section 3.10 of Gillespie 2004. But also see Gillespie 2006 for a discussion of why the rate of adaptive evolution might not depend on population size.

**106: Population size also influences the standing variation.** In addition to the effects of large population size on genetic variation discussed in this paragraph, larger populations tend to inhabit larger geographic areas than do smaller populations. Sub-populations may experience local adaptation, which can increase genetic diversity across the species as a whole.

**106: One important result of this increased efficiency of selection is that deleterious mutations under mutation–selection balance are kept at a low frequency.** However, more efficient selection in larger populations may also increase genetic diversity through heterozygote advantage and through frequency-dependent selection, when a mutation is advantageous if rare but disadvantageous if common. See Chapter 6 of Hartl and Clark 1997.

**106: previously neutral or deleterious alleles may become selectively advantageous.** This is called the Dykhuizen–Hartl effect (Dykhuizen and Hartl 1980). See also Gillespie 1973, Gillespie 1974a, Gillespie 1974b, Gillespie and Langley 1976, and Gillespie 1991 for models of evolution in fluctuating environments and Mustonen and Lassig 2009 for a new approach to studying evolution in variable environments.

**107: Figure 6.1.** Data on historical population growth from US Census Bureau (http://www.census.gov/ipc/www/worldhis.html and http://www.census.gov/ipc/www/idb/worldpop.html).

**108: Figure 6.2.** Adapted from MacLuliich 1937. For a review, see Korpimäki and Krebs 1996 and Krebs et al. 2001, and for more recent work on this problem, see Stenseth et al. 1997.

NOTES

**108: the long-term effective population size is biased strongly down-ward by the generations of reduced population size.** The long-term effective population size is approximated by the harmonic mean of population sizes across generations,

$$\frac{T}{\frac{1}{N_1}+\frac{1}{N_2}\cdots\frac{1}{N_T}},$$

where $N$ is the population size in each generation and $T$ is the number of generations (Crow and Kimura 1970; Hartl and Clark 1997).

**109: Figure 6.3.** Adapted from Grant and Grant 1993.

**109: The reduced effective population size of populations that fluctuate in size is expected to reduce the efficiency of natural selection.** Regular fluctuations in population size reduce the probability of fixation for a new allele (Ewens 1967). However, the probability of fixation in a fluctuating population is actually expected to fluctuate with the increases and decreases in population size. The exact dynamics are dependent on the relative strength of selection for a new mutation and the period of the population-size fluctuations (Otto and Whitlock 1997). Dramatic temporary reductions in population size, so-called population bottlenecks, may also play several roles in evolution. First, extreme population bottlenecks have probably occurred during the founding of new island populations. It is thought that many species of *Drosophila* endemic to the Hawaiian Islands arose from populations that were founded by a small number of emigrants from a neighboring island (Kaneshiro 1988). These founding populations would have possessed a small fraction of the total variation in the original population. By chance, the new populations may have contained novel combinations of epistatic loci, leading to novel patterns of evolution (Carson and Templeton 1984 and Goodnight 1987). Second, population bottlenecks may "transform" epistatic genetic variance into additive genetic variance. During population bottlenecks, allelic variation contributing to epistatic variance tends to be lost. Since the epistatic alleles contribute less to the overall genetic variance, the additive component of the genetic variance is increased (Cheverud and Routman 1995). But see Turelli and Barton 2006 for an argument that this "transformation" of epistatic into additive genetic variance results from models that incorporate unrealistic biological assumptions.

**109: When migrants only rarely make their way between the subpopulations, then the subpopulations will tend to behave like small populations.** The consequences of population subdivision depend very much on the number of migrants that move between populations each generation. A rather small number of migrants, on the order of one migrant every other generation, may be sufficient to homogenize allele frequencies for neutral loci among subpopulations. This result comes from Sewall Wright's island model (Wright 1951) and is clearly presented in section 5.5 of Gillespie 2004. See also sections 6.5 and 6.6 of Crow and Kimura 1970. Population subdivision usually causes a decrease in the effective population size relative to a panmictic population (Whitlock and Barton 1997).

205

**109: subpopulations may thus experience local fixation of globally dele-terious mutations by genetic drift.** Population structure also can increase the probability of fixation of deleterious alleles due to reduced effective population size (Whitlock 2003).

**109–110: If the species continues to occupy diverse habitats, then these locally adapted alleles are unlikely to spread through the entire popula-tion.** Variation in selective values across subpopulations adds considerable com-plexity to estimates of the fixation probability (Whitlock and Gomulkiewicz 2005 and Vuilleumier et al. 2008). Alleles that are beneficial locally, but deleterious glob-ally, are unlikely to fix deterministically but may fix by genetic drift. Alleles that are beneficial in some areas, deleterious in others, but on average advantageous, may fix with a higher probability than they would in a panmictic population.

**110: There is no evidence that any of these mutations are universally favorable for individuals of this species.** On the contrary, there is growing evidence that these mutations provide an advantage under only very local and specific conditions (Korves et al. 2007, Scarcelli et al. 2007, and Wilczek et al. 2009).

**110: Human subpopulations may have experienced local adaptation of alleles of the *Melanocortin 1 Receptor* gene.** Several studies have tested for pos-itive selection on *Melanocortin 1 Receptor* alleles in human populations. Most stud-ies have concluded that purifying selection acts in African populations and that selection has been relaxed in populations outside of Africa (Rana et al. 1999, Harding et al. 2000, and John et al. 2003). A more recent study has suggested that positive selection may have acted on alleles in European populations (Savage et al. 2008).

**110: Four different mutations that alter the amino acid sequence of the Melanocortin 1 Receptor protein are associated with red hair or light skin color in humans.** A valine to leucine substitution at residue 60 is associated with fair skin and blonde or light brown hair (Box et al. 1997). An aspartic acid to histidine substitution at residue 294 generates a recessive red-hair allele (Box et al. 1997). An arginine to cysteine substitution at residue 151 generates a recessive red-hair allele (Box et al. 1997, Frandberg et al. 1998, and Smith et al. 1998). An arginine to tryptophan substitution at residue 160 generates a recessive red-hair allele (Box et al. 1997 and Smith et al. 1998).

**112: The existence of multiple extinct hominine lineages suggests that we should not reject such a model out of hand.** Fossil evidence suggests that multiple hominine species lived contemporaneously (Wood 1996 and Carroll 2003).

**112: These may not be the best mutations for the job. They may, for example, have pleiotropic effects.** Otto 2004 examined the effect of pleiotropy on the probability of fixation. In her model, selection acts on a focal trait and the pleiotropic effects of a mutation have a net negative effect on fitness.

The pleiotropic effects tend to reduce the effective selection on the trait and thus reduce the probability of fixation of the mutation.

**112: If selection persists, then mutations of ever smaller effect will be favored** (Nijhout and Paulsen 1997, Orr 1998, and Gillespie 2006).

**113: Over the past 3 million years or so, approximately 100,000-year cycles of glaciation altered the available ecological niches dramatically and repeatedly on time scales similar to the time scales required to fix positively selected mutations.** See Hewitt 2000 for a review of the genetic effects of the Quaternary ice ages. The average time to fixation of a mutation was found by Kimura and Ohta 1969 and is dependent on both the effective population size and the selection coefficient. To substitute, an advantageous allele must replace all copies of the disadvantageous allele, which obviously takes more generations in a larger population. With increasing intensity of selection, fewer generations are required for the advantageous allele to substitute. A new neutral mutation that ultimately fixes takes approximately $4N_e$ generations to fix. Selectively advantageous mutations fix faster than this. Populations with effective population sizes on the order of thousands to millions are therefore expected to fix new adaptive mutations on time scales similar to the time scales of some major ecological disruptions. See Mustonen and Lassig 2009 for a discussion of the effects of selection on different scales.

**113: El Niño events create wild swings in the strength and direction of natural selection within the lifetimes of individual birds** (Grant and Grant 2002). For a gentle introduction, see Grant and Grant 2008.

**114: Figure 6.5.** Adapted from Grant and Grant 2002.

**114: Presumably, mutations of relatively large effect on body size and on beak morphology are being buffeted to higher frequencies, then to lower frequencies, then back again on the time scale of decades.** Fluctuating selection can have complicated effects on population genetics and on molecular evolution. Huerta-Sanchez et al. 2008 have examined effects of fluctuating selection coefficients on patterns of polymorphism and divergence. See Gillespie and Langley 1976, Gillespie 2006, Mustonen and Lassig 2007, and Mustonen and Lassig 2009 for additional discussion of the effects of fluctuating selection.

**115: The Heat Shock Factor protein, for example, forms a trimer upon exposure to heat and this trimer then activates transcription of many genes that protect the cell from heat damage** (Morimoto 1993).

**115: after exposure to cold days, transcription of the Flowering Locus C protein in *Arabidopsis thaliana* is downregulated, which allows the induction of flowering** (Michaels and Amasino 1999, Sheldon et al. 1999, Sheldon et al. 2000, and Simpson and Dean 2002).

**115: the mechanisms regulating plasticity are likely to permeate every aspect of the organism.** For book-length discussions of this topic, see Pigliucci 2001, West-Eberhard 2003, and DeWitt and Scheiner 2004. For shorter reviews, see Scheiner 1993, Pigliucci 1996, Nylin and Gotthard 1998, and Nijhout 1999.

## CHAPTER 7

**117: similar amino acid changes in Opsin proteins have evolved in different species to cause similar shifts in wavelength sensitivity** (Yokoyama and Yokoyama 1990, Yokoyama and Radlwimmer 1998, and Yokoyama 2002).

**117: the *cis*-regulatory region of the *lactase* gene has evolved multiple times to cause lactose tolerance** (Ingram et al. 2007 and Tishkoff et al. 2007). Some authors reserve the use of the term parallelism for cases in which identical nucleotide changes have occurred. I find this definition too restrictive and prefer to focus on parallelism at the level of genes, because this emphasizes how biochemical interactions within networks will evolve. For most purposes it is irrelevant whether lactase expression changed because of identical or different mutations in a *cis*-regulatory region.

**118–120: It also is useful to think of gene products as signals.** See Ptashne and Gann 1997 and Ptashne and Gann 2002 for discussion of this perspective.

**120: Figure 7.2.** The network in this figure, and many of the networks and pathworks shown in this chapter, were originally drawn using BioTapestry Editor software (Longabaugh et al. 2005 and Longabaugh et al. 2009), which is available free at http://www.biotapestry.org.

**121: The *IME1* gene appears to serve as a centralized decision point. It is the master regulator of sporulation.** To accommodate this decision-making role, *IME1* has evolved one of the largest *cis*-regulatory regions in yeast, of about 2000 bp (Granot et al. 1989 and Sagee et al. 1998). Other yeast genes that serve as master regulators of other developmental decisions also have relatively large *cis*-regulatory regions (Rupp et al. 1999).

**121: Most of the difference in sporulation efficiency between an oak-tree strain and a vineyard strain results from changes at the gene *RME1* and its transcriptional target, the *IME1* gene** (Gerke et al. 2009).

**121: Other changes in other genes also contribute to sporulation rate.** These polymorphisms also display extensive epistasis (Gerke et al. 2009). For example, the *RSF1* alleles cause less than a 3% change on their own, but their epistatic effects with alleles of *RME1* and *IME1* cause from 11% to 16% changes in sporulation rate.

**126: In a multicellular organism, different cells use different parts of the total regulatory network.** An analogous situation occurs in single-celled organisms when single genes participate in developmental decisions at different times in the life cycle. The principles discussed in this chapter that apply to networks in multicellular organisms also apply to temporal differences in single-celled organisms. For example, a gene that participates in only a single developmental decision throughout the life cycle will have fewer pleiotropic roles than a gene that participates in multiple developmental decisions at different times in the life cycle—and, of course, the lineage of a single cell in a multicellular organism can participate in different developmental decisions at different times. A thorough accounting of the pleiotropic roles of a gene must incorporate both spatial and temporal aspects of gene activity.

**128: Figure 7.3.** Adapted from Carroll et al. 2001.

**129–130: We do not necessarily know where we are headed, although following a random path will certainly take us somewhere.** There is an old saying, variously attributed to Lewis Carroll, George Harrison, and random rabbis: "If you don't know where you're going, all roads will take you there."

**130: The concept of pathworks provides many advantages.** Pathworks also elucidate what might otherwise be considered a surprising fact: that developmental genetics, as a discipline, works. If all genes participated in complex networks of interactions in every cell of the body, all of the time, then we would not be able to dissect the contribution of individual genes to development. We would not be able to eliminate the function of a single gene and observe specific phenotypic defects. Instead, we would observe catastrophic failure of the system whenever we disrupted any part of it. This, in turn, raises the question of why development has evolved to utilize pathworks. For one thing, pathworks generate modularity and modularity may enhance evolvability (Gerhart and Kirschner 1997).

**131: Figure 7.5.** The drawing of a first-instar larvae of *Drosophila melanogaster* comes from http://Flybase.org. The photomicrograph of larval trichomes was kindly provided by François Payre.

**132: Current evidence indicates that at least some of these genes are directly regulated by the Shavenbaby protein** (Chanut-Delalande et al. 2006).

**132: Transcription of the *shavenbaby* gene was upregulated by the activity of . . . the *SoxNeuro* and *Dichaete* genes** (Overton et al. 2007).

**133: The signal from the EGF Receptor pathway emanated from a cell just to the posterior of our focal cell . . . Production of this signal required expression of the Rhomboid protein in this more posterior cell** (Walters et al. 2005).

**133: Transcription of the *rhomboid* gene was activated in this cell by positive signals from the Hedgehog signaling pathway** (Alexandre et al. 1999).

**133: Transcription of the *hedgehog* gene in this cell was activated by the Engrailed protein** (Hatini and DiNardo 2001).

**137: Current evidence suggests that changes in the expression pattern of the *shavenbaby* gene caused these changes in trichome patterns in the *Drosophila virilis* clade.** Three pieces of evidence suggest, but do not yet prove, that changes in the *shavenbaby* gene have caused trichome evolution in the

*Drosophila virilis* clade (Sucena et al. 2003). First, an earlier study of some of the genes that act upstream of the *shavenbaby* gene showed that none of these upstream regulators have evolved new expression patterns in the *Drosophila virilis* group (Dickinson et al. 1993). Second, interspecific genetic analysis implicated one or more genetic factors on the X chromosome as responsible entirely for the difference in trichome pattern between several species. The *shavenbaby* gene resides on the X chromosome in these species. Third, the expression pattern of the Shavenbaby mRNA precisely matched the trichome patterns produced by nine species of the *Drosophila virilis* group. Further studies of the *cis*-regulatory region of the *shavenbaby* gene are required to determine whether the evolutionary changes occurred in the *shavenbaby* gene or in one or more genes upstream of the *shavenbaby* gene.

**137: the olive fruit fly . . . has evolved a pattern of trichomes . . . [that] is precisely correlated with expression of the Shavenbaby gene product** (Khila et al. 2003).

**139: The *shavenbaby* gene appears to be a hot spot for evolutionary changes in larval trichome pattern.** Current evidence indicates that either the *shavenbaby* gene or one or more genes upstream of the *shavenbaby* gene are likely to be responsible for the multiple evolutionary changes in trichome patterns discussed here. This is similar to the yeast sporulation pathway, where the master regulator for sporulation—the *IME1* gene—and upstream components have evolved.

**140: Changes in individual genes downstream of the Shavenbaby protein can cause slight changes in the shape of a trichome, but blocking the activity of any single downstream gene does not prevent differentiation of a trichome** (Chanut-Delalande et al. 2006).

**142: Figure 7.11 illustrates much of the genetic network that transduces these signals to regulate flowering time.** The network in Figure 7.11 was adapted, with help from Caroline Dean and David Laurie, from the following references: Levy and Dean 1998, Simpson and Dean 2002, Imaizumi and Kay 2006, and Jung et al. 2007.

**144: two known mutations of the *Flowering Locus C* gene in *Arabidopsis thaliana* populations . . . alter vernalization** (Gazzani et al. 2003 and Michaels et al. 2003).

**144: Both mutations alter the *cis*-regulatory region of the *Flowering Locus C* gene.** There are several genes in the *Arabidopsis thaliana* genome that are closely related to the *Flowering Locus C* gene. At least one of them, the *Flowering Locus M* gene, also regulates vernalization (Ratcliffe et al. 2001 and Scortecci et al. 2001). The precise role of the *Flowering Locus M* gene is not well understood, but it appears to act downstream of, or independently of, the *Flowering Locus C* gene (Ratcliffe et al. 2001). A deletion mutation of the *Flowering Locus M* gene contributes to natural variation in flowering time in *Arabidopsis thaliana* (Werner et al. 2005).

**145: null mutations of the *Frigida* gene do have pleiotropic effects on growth rate** (Korves et al. 2007 and Scarcelli et al. 2007).

**147: Development biases genetic evolution towards hot-spot genes.** This is different from saying that development biases phenotypic evolution, a claim for which I believe the evidence is less compelling. My claim is simply that, given a particular selective pressure, mutations in genes that integrate information and control a module of development are more likely to be fixed than mutations in genes found in other locations in the pathwork. However, see Kavanagh et al. 2007 for perhaps the best example of how development may bias phenotypic evolution. For another take on the idea that development biases genetic evolution, see Wilkins 2007.

## CHAPTER 8

**150: to the extent that genetic evolution is sensitive to dominance, pleiotropy, and epistasis, substitutions will occur preferentially in specific, and predictable, locations in pathworks and in genes.** Starting in 1954, Ernst Mayr published a set of ideas that may seem superficially similar to the idea that evolution favors mutations with predictable effects (see Mayr 1954 and Mayr 1963). Mayr called some genes "good mixers" and integrated this into his somewhat slippery view of the "cohesion of the gene pool." Mayr was searching for a genetic explanation for the origin of new species. He recognized that populations that exchanged alleles would have little chance of separating into reproductively isolated populations. He was particularly impressed with genetic evidence for epistasis, and he believed that populations of a single species represented a cohesive genetic whole:

> The essence of speciation, as we now realize, is the production of two well-integrated gene complexes from a single parental one. All early attempts to explain the genetics of speciation missed this essential point, being concerned entirely with the problem of the origin of difference. . . . The real problem of speciation is not how to produce difference but rather how to escape from the cohesion of the gene complex. No one will comprehend how formidable this problem is who does not understand the power of the cohesive forces that are responsible for the coadapted harmony of the gene pool [Mayr, 1963 p. 518].

Mayr devoted an entire chapter of his 1963 book *Animal Species and Evolution* to describing "cohesion" of the gene pool. Mayr knew that natural populations contain considerable genetic variation for fitness, and he was also aware that much of this variation resulted from epistasis and balancing selection. He interpreted epistasis and balancing selection as the "cohesive" forces of populations that stabilized the phenotype in response to the new input of genetic variation. In particular, Mayr argued that the existence of this variation in large populations would select for new alleles that were "good mixers." What Mayr meant by "good mixers" was alleles that "produce viable heterozygotes with a great assortment of alien genes or gene combinations" (Mayr 1954). Mayr clarified this idea in a figure from his 1954 paper, where he illustrated the adaptive value of two alleles in different genetic backgrounds, one a good mixer (a2, the heavy line) and one a poor mixer (a1, the light line).

Differing genetic backgrounds

While Mayr implied that a good mixer allele is good because it has favorable epistatic interactions in many different genetic backgrounds, a more straightforward interpretation of his figure is that good mixers are alleles with minimal epistatic interactions and with stronger additive effects. In addition, since Mayr expected these alleles to persist in the population to become a stable part of the gene pool, they must also display balancing selection. So a simpler interpretation of Mayr's "good mixers" is that they are alleles under balancing selection displaying minimal epistatic interactions.

I emphasize in Chapter 5 that new alleles with epistatic effects may exhibit higher variance in fitness than alleles without epistatic effects and, therefore, may have a lower probability of fixation (i.e. a higher probability of going extinct). Thus, Mayr's "good mixers" are fundamentally different from the kinds of epistatic interactions I describe in Chapter 5, and the term "good mixers" does not clarify the role of epistasis in evolution.

**152: Figure 8.1, selection on phenotypic traits in the wild shows an approximately exponential distribution.** These data come from Kingsolver et al. 2001. The intensity of selection shown in this figure, adapted from Hoekstra et al. 2001, is measured as the linear selection gradient (Phillips and Arnold 1989). The selection gradient estimates the intensity of selection on each trait independent of effects on correlated traits and is defined as the selection differential on a trait divided by the phenotypic standard deviation of the trait before selection. The selection differential is defined as the difference in the phenotypic mean of a population after and before an episode of selection. The selection gradient allows meaningful comparisons of the intensity of selection between traits and between studies (Falconer and Mackay 1996 and Roff 2003).

**154: if null alleles in developmental genes played an important role in developmental evolution, then we would observe few highly conserved developmental genes.** Null alleles do contribute to genome evolution (Blomme et al. 2006, Demuth et al. 2006, and Tanaka et al. 2006). Some lineages have experienced high gene turnover, with extensive gene gain and loss. High turnover seems to have been particularly important for some gene families, such as odorant receptors. Gene loss is much less frequent for genes involved in core cell-biological and developmental processes, although it does occur (Blomme et al. 2006, Lemons and McGinnis 2006, and Tanaka et al. 2006).

**155: we are still far from generating the kind of unbiased data set that is required to test models of genetic evolution.** An unbiased survey implies that the entire genome was examined for mutations contributing to a phenotypic difference. In practice, this means that phenotypic differences should be genetically mapped first with a whole-genome scan and then with high resolution mapping of targeted regions that contain the relevant mutations. It has become common to jump to study candidate genes. While this sometimes leads to discovery of relevant mutations, this logical jump introduces bias and makes it difficult to use these data for comparative purposes.

**155: More than 300 mutations causing phenotypic variation in domesticated races, within species, and between species have been reported.** These data were collated and reviewed initially by Stern and Orgogozo 2008. The article, data, and criteria used to include studies are available at http://www3 .interscience.wiley.com/cgi-bin/fulltext/120696379/HTMLSTART

**155: Many of these mutations were identified by studies that focused on candidate genes, and this fact, along with many others that I discuss at length in the Notes section at the end of this book, render the body of existing data a highly suspect substrate for testing models of genetic evolution.** There are at least six reasons to be suspicious of the data discussed in the survey of genetic evolution in natural populations.

1. These are "found data," and they are reported in heterogeneous studies with different standards of evidence. These data do not represent a random sample of the mutations contributing to evolution of any particular phenotypic trait. Most traits have been selected because they are of interest to the investigators and differ greatly between lines or species. Most of the traits studied in domesticated species are traits of economic importance.
2. Investigators in different fields, such as physiology and development, have different expectations about where relevant mutations will be found. It is likely that physiologists are biased to believe that relevant mutations will be found in the coding regions, whereas many developmental biologists are biased to suspect that the relevant mutations will be found in *cis*-regulatory regions.
3. The studies generating the data reviewed in this chapter were biased by the tools available for discovery of the relevant mutations. Most studies are biased toward discovering coding, versus *cis*-regulatory, mutations because it is much easier to discover coding mutations than *cis*-regulatory mutations in most systems. Many investigators explicitly note that they first examined the coding region of genes and, if no obviously relevant mutations were found, they then began to look in the *cis*-regulatory region.

4. Many studies represent resampling of the same gene in different species after the initial discovery of mutations causing phenotypic variation. This is particularly true for candidate genes, such as the *Melanocortin 1 Receptor* gene, where it is relatively easy to infer the phenotypic consequences of molecular variation.

5. The genes underlying quantitative traits—traits determined by variation at multiple genes—often have variable effect sizes. For practical reasons, the mutations at genes causing the largest effects are normally discovered first. Mutations causing large phenotypic effects may cause different kinds of molecular changes than mutations causing weak phenotypic effects.

6. These studies include an excess of candidate genes. In many cases, investigators studied candidate genes identified from earlier functional studies. In other cases, investigators used a hybrid approach, performing unbiased genetic mapping to localize the general genomic region containing evolutionarily relevant variation and then identifying candidate genes within this region for further study. It is not yet clear whether this second, biased step provides an accurate view of the genetic causes of phenotypic variation. Some investigators have taken a third approach: they first identify an interesting gene and then try to identify the phenotypic consequences of molecular changes in this gene. For some genes, like the hemoglobin genes, this is reasonable. For genes with poorly characterized biochemical roles, the inference of the phenotypic consequences of molecular variation is fraught with complications.

**155: these trends provide a strong impetus to collect more data.** The best data set would result from unbiased collection of mutations contributing to phenotypic evolution.

**155–156: coding changes are far easier to discover than are *cis*-regulatory changes.** This results from several facts. First, coding regions are usually smaller than are *cis*-regulatory regions. Second, coding regions are defined clearly by the triplet genetic code, whereas *cis*-regulatory regions do not display such strong structure. Third, we possess myriad techniques for testing changes in protein function and relatively few reliable techniques for testing function of *cis*-regulatory regions. Fourth, changes in *cis*-regulatory regions can cause changes in quantitative levels of expression or changes in spatial or temporal patterns of expression. It is not always obvious which change has occurred, and sometimes it is challenging to assay quantitative spatial and temporal changes in expression.

**156: null mutations frequently cause phenotypic variation both within populations (25%) and in domesticated races (56%), whereas only 7% of the mutations causing species differences are null mutations.** This is a statistically significant difference (Two-tailed Fisher's Exact test, $P = 1.5 \times 10^{-12}$).

**157: a transposable element insertion upstream from a gene encoding a Cytochrome P450 protein in *Drosophila melanogaster* causes increased gene expression in specific tissues and confers insecticide resistance** (Chung et al. 2007).

**157: large insertion or deletion events are implicated in zero out of 27 cases of *cis*-regulatory evolution between species.** This result is discussed further in Stern and Orgogozo 2008. The difference in the frequency of large insertions and deletions in *cis*-regulatory regions within species versus between species is statistically significant (Two-tailed Fisher's Exact test, $P = 0.003$).

**157–158: flies carrying large insertions in a 106,000 bp region encompassing the *achaete* and *scute* genes . . . had an average of 1.62 fewer sternopleural bristles and 1.18 fewer abdominal bristles than did flies without these insertions** (Mackay and Langley 1990).

**158: One of these insertions . . . was found at a frequency of 14%, suggesting that it is not universally deleterious and may even be maintained . . . by balancing selection** (Long et al. 2000).

**158: The excess of *cis*-regulatory mutations causing morphological evolution between species is consistent with theoretical arguments discussed in Chapter 4.** Wittkopp et al. 2008 found that the causes of variation in gene expression levels between species are biased toward *cis*-regulatory changes. This provides evidence independent of the kinds of studies that I have reviewed that genetic differences between species are biased towards *cis*-regulatory changes.

**159: It is not entirely clear why evolution of developmental and physiological genes would differ in this way.** Some have suggested that we should expect physiological evolution to involve less *cis*-regulatory evolution because genes involved in physiology have smaller *cis*-regulatory regions than do developmental genes (Carroll 2008). However, there is also evidence for, and theoretical arguments supporting, a larger role for *cis*-regulatory changes than for protein coding changes during physiological evolution (Hoekstra and Coyne 2007 and Rosati et al. 2008).

**162: Experimental evolution tends to be performed on microorganisms . . . that reproduce asexually. Typically, the experiment is initiated with a single clone.** See Elena and Lenski 2003 for a review of experimental evolution studies in microorganisms. Experimental evolution also has been performed starting with a heterogeneous mixture of clones (de Visser and Rozen 2006 and Kao and Sherlock 2008) and with sexual organisms. See Chippindale 2006 for a review of experimental evolution studies in sexual species.

**162: New adaptive mutations often sweep through the population, and adaptive evolution usually involves a series of consecutive selective sweeps.** When the results of experimental evolution are studied with sufficient temporal resolution, steplike increases in fitness can be observed (Lenski and Travisano 1994 and Burch and Chao 1999). This observation implies that mutations intermittently sweep through populations. In large asexual populations, it is also possible for a second adaptive mutation to arise in a population while a first adaptive mutation is heading to fixation. The different mutations may then compete, a phenomenon called "clonal interference." Clonal interference can slow the fixation of, and can even cause the elimination of, the first mutation (Fisher 1930, Muller 1932, Gerrish and Lenski 1998, de Visser and Rozen 2006, and Kao and Sherlock 2008).

**163: Replicate experimental–evolution populations often evolve similar or identical genetic changes.** In addition to the simple mutations discussed in this paragraph, clonal microorganisms experiencing experimental evolution often evolve dramatic genomic changes, including large deletions and genome rearrangements (Riehle et al. 2001 and Dunham et al. 2002).

**163: when the bacteriophage $\phi$X 174 was reared at a high temperature in two replicate populations in a novel host, *Salmonella typhimurium*, half of the same codons mutated to the same amino acid in both replicates** (Wichman et al. 1999).

**163: experimental evolution provides compelling evidence for parallel evolution at the nucleotide level.** We expect to observe similar molecular convergence only rarely in multicellular organisms, which have much larger genome sizes than does this bacteriophage. The small genome size of this bacteriophage, together with the large population sizes and numerous generations of experimental evolution, increased the odds that similar mutations would arise in replicate populations. For example, the human immunodeficiency virus, which contains a small genome, experienced evolution of precisely the same amino-acid changes in different hosts when challenged with antiretroviral drugs (Boucher et al. 1992, Kellam et al. 1994, and Borman et al. 1996).

**163: new mutations contribute to an increasing fraction of the response to selection after about the first 20 generations** (Barton and Keightley 2002).

**164: By and large, this variation will have been either neutral or under balancing selection and likely to have been either formerly segregating at intermediate frequency or deleterious and present in the original population at low frequency** (Simmons and Crow 1977).

**164: Artificial selection experiments result often in selection of variants with epistatic and perhaps pleiotropic effects.** Data are reviewed in Barton and Keightley 2002 and in Mackay 2004.

**164: selection for increased and decreased numbers of bristles . . . resulted in a strong evolutionary change in abdominal bristle number** (Long et al. 1995).

**164: Many of these alleles also changed the number of bristles on the thorax, suggesting that these alleles had pleiotropic effects.** An alternative explanation is that the apparent pleiotropic effects are due to alleles at closely linked loci that were not separated from the selected alleles during the short course of the experiment. Similarly, the reduced viability attributed to some loci may have resulted from selection of linked loci. During a short bout of strong selection, tightly linked mutations would tend to act like a single allele, giving the appearance of pleiotropy.

**165: Figure 8.4.** The drawing of a *Drosophila melanogaster* ventral abdomen comes from http://Flybase.org. The graph of the evolution of bristle numbers was adapted from Long et al. 1995.

**165: selection on abdominal bristle numbers for more than 80 generations led to a dramatic increase in bristle number in six replicate lines** (Yoo 1980b).

**165–166: The increases in bristle number in later generations resulted largely from selection for recessive lethal mutations and for visible mutations of large effect that arose during the experiment** (Yoo 1980a).

**167: Figure 8.5.** The photograph is from McPherron and Lee 1997 and is used with permission. © 1997 by the National Academy of Sciences, USA.

**167: The mutation that causes wrinkled pea seeds . . . is a transposable-element insertion that inactivates a gene producing a starch-branching enzyme** (Mendel 1865 and Bateson 1901).

**167: Six different null mutations in the *myostatin* gene cause muscle hypertrophy in different breeds of cattle** (Grobet et al. 1997, Kambadur et al. 1997, McPherron and Lee 1997, and Grobet et al. 1998).

**167: A null mutation in the *myostatin* gene also causes muscle hypertrophy in some individuals of the whippet dog breed** (Mosher et al. 2007).

**168: White grapes . . . are caused by insertion of a transposable element upstream of the *VvmybA1* gene** (Kobayashi et al. 2004).

**168: A single base pair change in the *cis*-regulatory region of the *insulin-like growth factor 2* gene controls 15–30% of the difference in muscle mass** (Van Laere et al. 2003).

**168: This fundamental difference in plant architecture was caused almost entirely by evolution of the *teosinte branched1* gene** (Doebley et al. 1997).

**169: Figure 8.6.** Figure from Doebley et al. 1995 and the original figure was kindly provided by John Doebley.

**169: The higher levels of Teosinte branched1 transcript result from one or more mutations in a *cis*-regulatory region far upstream of the first exon of the *teosinte branched1* gene** (Clark et al. 2006).

**170: Amongst the assumptions most likely to be suspect and problematic are assumptions of constant selection coefficients and of constant population sizes.** Fluctuating selection coefficients can affect the probability of fixation of a new mutation (Gillespie 1973, Gillespie 1974c, Gillespie, 1975, Takahata et al. 1975, and Gillespie 1977), as can fluctuating population sizes (Otto and Whitlock 1997), heterogeneous environments (Gillespie 1974a, Gillespie 1974b, Gillespie and Langley 1976, and Whitlock and Gomulkiewicz 2005), and subdivided populations (Whitlock 2003 and Vuilleumier et al. 2008). See also Musto-nen and Lassig 2009 for a discussion of genomic evolution in response to selection acting on different time scales.

**170–171: Darwin famously wrote, in** *The Origin of Species*, **"If it could be demonstrated that any complex organ existed, which could not possibly have been formed by numerous, successive, slight modifications, my theory would absolutely break down."** See p. 189 of Darwin 1859. This book is available free online at http://darwin-online.org.uk/EditorialIntroductions/Freeman_OntheOriginofSpecies.html.

**173: Perhaps—just perhaps—while most populations evolve obediently through the fixation of mutations with specific effects, small popula-tions, while teetering toward extinction and irrelevance, provide caul-drons of evolutionary novelty.** Many times over the past century, different biologists have promoted the importance of small population sizes or population bottlenecks for evolution. Wright argued that spatially structured subpopulations within a species would experience significant genetic drift and allow favorable epistatic combinations of alleles to spread throughout the species (Wright 1931, Wright 1955, and Wright 1980). Others have argued that peripherally isolated subpopulations and population bottlenecks can lead to a "genetic revolution" (Mayr 1954) and thereby promote speciation (Carson and Templeton 1984). In a similar vein, some authors have argued that reduced population sizes can promote rapid evolution, perhaps even generating macroevolutionary patterns such as punctuated equilibrium (Eldredge 1972, Gould and Eldredge 1977, and Gould and Eldredge 1986).

## HISTORICAL NOTE

**176: Figure HN.1.** Reproduced from Figure 19 of Bateson 1894.

**176:** *The Material Basis of Evolution* (Goldschmidt 1940).

**177: most populations harbor extensive reserves of genetic variation consisting mostly of mutations of small effect.** See Mackay 2001 and Mackay 2004 for reviews of this topic.

**177: closely related species tend to differ in small and quantitative ways from each other.** Probably the easiest way to see this is to pick up any taxo-nomic treatise on a group of organisms or to visit a good museum collection. Many textbooks on evolution also illustrate some of the variation found in nature. Also see Falconer and Mackay 1996.

**177: distantly related species display obvious discontinuities—jumps in phenotypic space.** This is one of the observations that motivated Bateson's view that natural selection provided an insufficient explanation for evolution and why he turned to Mendelian genetic factors of large effect as a better explanation for evolution between species. Read his Introduction to *Materials for the Study of Varia-*

*tion* (Bateson 1894). This observation also motivated Goldschmidt's thinking (Goldschmidt 1940).

**178: Figure HN.2** Photograph of bithorax fly by Ed Lewis. Photographs of the butterfly *Idea idea* and the dragonfly *Epiaeschna heros* are from http:// butterflyutopia.com and kindly provided by Paul Caparatta and used with his permission.

**178: in all insects examined so far, expression of Ultrabithorax protein in the hind wings causes the hind wings to look different from the forewings.** Ultrabithorax protein is expressed in the hind wings of butterflies (Weatherbee et al. 1999) and beetles (Tomoyasu et al. 2005). Ultrabithorax protein is expressed in the third pair of legs, and possibly also in the hind wings, of locusts and crickets (Kelsh et al. 1994 and Mahfooz et al. 2007), and milkweed bugs (Angelini et al. 2005 and Mahfooz et al. 2007). Ultrabithorax protein regulates multiple genes to influence the morphology of the hind wings (Weatherbee et al. 1998 and Roch and Akam 2000).

**179: the expression patterns of these genes correlate with dramatic differences in body plans between distantly related taxa.** See, for example, Averof and Patel 1997 and Cohn and Tickle 1999.

# FURTHER READING

CHAPTER 1

*Levels of analysis*
Mayr, E. 1961. Cause and effect in biology. *Science* 134:1501–1506.

Frigida
Levy, Y. Y., and C. Dean. 1998. The transition to flowering. *The Plant Cell* 10: 1973–1990.
Simpson, G. G., and C. Dean. 2002. *Arabidopsis*, the Rosetta stone of flowering time? *Science* 296:285–289.

CHAPTER 2

*Population genetics*
Hartl, D. L., and A. G. Clark. 1997. *Principles of Population Genetics*. Sinauer Associates, Sunderland, MA.
Gillespie, J. H. 2004. *Population Genetics: A Concise Guide*. The Johns Hopkins University Press, Baltimore, MD.

*Molecular evolution*
Li, W.-H. 1997. *Molecular Evolution*. Sinauer Associates, Sunderland, MA.

*Gene function and evolution*
Lovell, S. C. 2006. Gene function and molecular evolution. Pp. 193–210 *in* C. W. Fox and J. B. Wolf, eds. *Evolutionary Genetics: Concepts and Case Studies*. Oxford University Press, Oxford.

*Transcription-factor function and evolution*
Carroll, S. B., J. K. Grenier, and S. D. Weatherbee. 2001. *From DNA to Diversity: Molecular Genetics and the Evolution of Animal Design, 2nd Edition*. Blackwell Publishing, Malden, MA.
Carroll, S. B. 2005. Evolution at two levels: on genes and form. *PLoS Biology* 3:e245.
Lynch, V. J., and G. P. Wagner. 2008. Resurrecting the role of transcription factor change in developmental evolution. *Evolution* 62:2131–2154.

## CHAPTER 3

### Physiological dominance

Wright, S. 1934. Physiological and evolutionary theories of dominance. *The American Naturalist* 68:24–53.

Kacser, H., and J. A. Burns. 1981. The molecular basis of dominance. *Genetics* 97:639–666.

### Transcriptional regulation

White, R. J. 2001. *Gene Transcription: Mechanisms and Control.* Blackwell Science Ltd., Oxford.

Davidson, E. H. 2001. *Genomic Regulatory Systems.* Academic Press, San Diego, CA.

Davidson, E. H. 2006. *The Regulatory Genome: Gene Regulatory Networks in Development and Evolution.* Academic Press, Burlington, VT.

Ptashne, M., and A. Gann. 2002. *Genes & Signals.* Cold Spring Harbor Laboratory Press, Cold Spring Harbor, NY.

### Drosophila development

Lawrence, P. A. 1992. *The Making of a Fly.* Blackwell Scientific Publications, Oxford.

### Probability of fixation of mutation

Crow, J. E., and M. Kimura. 1970. *An Introduction to Population Genetics Theory.* Harper & Row Publishers, New York.

Gillespie, J. H. 2004. *Population Genetics: A Concise Guide.* The Johns Hopkins University Press, Baltimore, MD.

Gillespie, J. H. 2006. Stochastic processes in evolution. Pp. 65–79 *in* C. W. Fox and J. B. Wolf, eds. *Evolutionary Genetics: Concepts and Case Studies.* Oxford University Press, Oxford.

### Mutation–selection balance

Hartl, D. L., and A. G. Clark. 1997. *Principles of Population Genetics.* Sinauer Press, Sunderland, MA.

### Haldane's sieve

Orr, H. A., and A. J. Betancourt. 2001. Haldane's sieve and adaptation from the standing genetic variation. *Genetics* 157:875–884.

## CHAPTER 4

### Evolution of cis-regulatory DNA

Wray, G. A. 2007. The evolutionary significance of *cis*-regulatory mutations. *Nature Reviews Genetics* 8:206–216.

Wray, G. A., M. W. Hahn, E. Abouheif, J. P. Balhoff, M. Pizer, M. V. Rockman, and L. A. Romano. 2003. The evolution of transcriptional regulation in eukaryotes. *Molecular Biology and Evolution* 20:1377–1419.

Carroll, S. B. 2005. Evolution at two levels: on genes and form. *PLoS Biology* 3:e245.

Carroll, S. B. 2008. Evo-devo and an expanding evolutionary synthesis: a genetic theory of morphological evolution. *Cell* 134:25–36.

*Pigmentation evolution*

Hoekstra, H. E. 2006. Genetics, development and evolution of adaptive pigmentation in vertebrates. *Heredity* 97:222–234.

Wittkopp, P. J., and P. Beldade. 2009. Development and evolution of insect pigmentation: genetic mechanisms and the potential consequences of pleiotropy. *Seminars in Cell and Developmental Biology* 20:65–71.

*Pleiotropy and Evolution*

Otto, S. P. 2004. Two steps forward, one step back: the pleiotropic effects of favoured alleles. *Proceedings of the Royal Society B: Biological Sciences* 271:705–714.

Stern, D. L. 2000. Perspective: Evolutionary developmental biology and the problem of variation. *Evolution* 54:1079–1091.

*Fisher's geometric model of adaptation*

Fisher, R. A. 1958. *The Genetical Theory of Natural Selection.* Dover, New York.

Orr, H. A. 1998. The population genetics of adaptation: the distribution of factors fixed during adaptive evolution. *Evolution* 52:935–949.

Barton, N. H., and P. D. Keightley. 2002. Understanding quantitative genetic variation. *Nature Reviews Genetics* 3:11–21.

*Gene duplication*

Ohno, S. 1970. *Evolution by Gene Duplication.* Springer-Verlag, New York.

Lynch, M., and J. S. Conery. 2000. The evolutionary fate and consequences of duplicate genes. *Science* 290:1151–1155.

Zhang, J. 2003. Evolution by gene duplication: an update. *Trends in Ecology & Evolution* 18:292–298.

## CHAPTER 5

*Genetic epistasis tests*

Avery, L., and S. Wasserman. 1992. Ordering gene function: the interpretation of epistasis in regulatory hierarchies. *Trends in Genetics* 8:312–316.

*Epistasis and evolution*

Wolf, J. B., E. D. Brodie III, and M. J. Wade, eds. 2000. *Epistasis and the Evolutionary Process.* Oxford University Press, Oxford.

Weinreich, D. M., R. A. Watson, and L. Chao. 2005. Perspective: Sign epistasis and genetic constraint on evolutionary trajectories. *Evolution* 59:1165–1174.

Moore, J. H., and S. M. Williams. 2005. Traversing the conceptual divide between biological and statistical epistasis: systems biology and a more modern synthesis. *Bioessays* 27:637–646.

Phillips, P. C. 2008. Epistasis—the essential role of gene interactions in the structure and evolution of genetic systems. *Nature Reviews Genetics* 9:855–867.

*Hidden genetic variation*

Gibson, G., and I. Dworkin. 2004. Uncovering cryptic genetic variation. *Nature Reviews Genetics* 5:681–690.


## CHAPTER 6

*Population size and structure and evolution*

Gillespie, J. H. 2006. Stochastic processes in evolution. Pp. 65–79 *in* C. W. Fox and J. B. Wolf, eds. *Evolutionary Genetics: Concepts and Case Studies*. Oxford University Press, Oxford.

Goodnight, C. J. 2006. Genetics and evolution in structured populations. Pp. 80–100 *in* C. W. Fox and J. B. Wolf, eds. *Evolutionary Genetics: Concepts and Case Studies*. Oxford University Press, Oxford.

Hanski, I. A., and O. E. Gaggitotti, eds. 2004. *Ecology, Genetics, and Evolution of Metapopulations*. Elsevier, Burlington, VT.

*Evolution in a changing environment*

Barrett, R. D., and D. Schluter. 2008. Adaptation from standing genetic variation. *Trends in Ecology and Evolution* 23:38–44.

*Phenotypic plasticity and evolution*

DeWitt, T. J., and S. M. Scheiner. 2004. *Phenotypic Plasticity: Functional and Conceptual Approaches*. Oxford University Press, Oxford.


## CHAPTER 7

*Genetic networks*

Davidson, E. H. 2001. *Genomic Regulatory Systems*. Academic Press, San Diego, CA.

Davidson, E. H. 2006. *The Regulatory Genome: Gene Regulatory Networks in Development and Evolution*. Academic Press, Burlington, VT.

Wilkins, A. S. 2002. *The Evolution of Developmental Pathways*. Sinauer Associates, Sunderland, MA.


## CHAPTER 8

*Predictability of genetic evolution*

Golding, G. B., and A. M. Dean. 1998. The structural basis of molecular adaptation. *Molecular Biology and Evolution* 15:355–369.

Carroll, S. B. 2008. Evo-devo and an expanding evolutionary synthesis: a genetic theory of morphological evolution. *Cell* 134:25–36.

Stern, D. L., and V. Orgogozo. 2009. Is genetic evolution predictable? *Science* 323:746–751.

*Experimental evolution*

Elena, S. F., and R. E. Lenski. 2003. Evolution experiments with microorganisms: the dynamics and genetic bases of adaptation. *Nature Reviews Genetics* 4:457–469.

Chippindale, A. K. 2006. Experimental evolution. Pp. 482–501 *in* C. W. Fox and J. B. Wolf, eds. *Evolutionary Genetics: Concepts and Case Studies.* Oxford University Press, Oxford.

*Domestication*

Andersson, L., and M. Georges. 2004. Domestic-animal genomics: deciphering the genetics of complex traits. *Nature Reviews Genetics* 5:202–212.

Doebley, J. F., B. S. Gaut, and B. D. Smith. 2006. The molecular genetics of crop domestication. *Cell* 127:1309–1321.

## HISTORICAL NOTE

*Macromutationism versus micromutationism*

Akam, M. 1998. *Hox* genes, homeosis and the evolution of segment identity: no need for hopeless monsters. *The International Journal of Developmental Biology* 42:445–451.

Orr, H. A., and J. A. Coyne. 1992. The genetics of adaptation: A reassessment. *The American Naturalist* 140:725–742.

Orr, H. A. 2001. The genetics of species differences. *Trends in Ecology and Evolution* 16:343–350.

# BIBLIOGRAPHY

Alexandre, C., M. Lecourtois, and J. Vincent. 1999. Wingless and Hedgehog pattern *Drosophila* denticle belts by regulating the production of short-range signals. *Development* 126:5689–5698.

Allison, A. C. 1954. Protection afforded by sickle-cell trait against subtertian malarial infection. *British Medical Journal* 1:290–294.

Andreev, D., M. Kreitman, T. W. Phillips, R. W. Beeman, and R. H. ffrench-Constant. 1999. Multiple origins of cyclodiene insecticide resistance in *Tribolium castaneum* (Coleoptera: Tenebrionidae). *Journal of Molecular Evolution* 48:615–624.

Angelini, D. R., P. Z. Liu, C. L. Hughes, and T. C. Kaufman. 2005. *Hox* gene function and interaction in the milkweed bug *Oncopeltus fasciatus* (Hemiptera). *Developmental Biology* 287:440–455.

Argeson, A. C., K. K. Nelson, and L. D. Siracusa. 1996. Molecular basis of the pleiotropic phenotype of mice carrying the *hypervariable yellow* ($A^{hvy}$) mutation at the *agouti* locus. *Genetics* 142:557–567.

Arthur, W. 1984. *Mechanisms of Morphological Evolution: A Combined Genetic, Developmental and Ecological Approach.* Wiley, Chichester, UK.

Averof, M., and N. H. Patel. 1997. Crustacean appendage evolution associated with changes in *Hox* gene expression. *Nature* 388:682–686.

Avery, L., and S. Wasserman. 1992. Ordering gene function: the interpretation of epistasis in regulatory hierarchies. *Trends in Genetics* 8:312–316.

Baatz, M., and G. P. Wagner. 1997. Adaptive inertia caused by hidden pleiotropic effects. *Theoretical Population Biology* 51:49–66.

Barrett, R. D., S. M. Rogers, and D. Schluter. 2008. Natural selection on a major armor gene in threespine stickleback. *Science* 322:255–257.

Barton, N. H. 1995. Linkage and the limits to natural selection. *Genetics* 140:821–841.

Barton, N. H., and P. D. Keightley. 2002. Understanding quantitative genetic variation. *Nature Reviews Genetics* 3:11–21.

Bastock, M. 1956. A gene mutation which changes a behavior pattern. *Evolution* 10:421–439.

Bateson, W. 1894. *Materials for the Study of Variation Treated with Especial Regard to Discontinuity in the Origin of Species.* Macmillan and Co., London.

Bateson, W. 1900. Hybridization and cross-breeding as a method of scientific investigation. *Journal of the Royal Horticultural Society* 24:59–66.

Bateson, W. 1901. Experiments in plant hybridization by Gregor Mendel. *Journal of the Royal Horticultural Society* 24:1–32.

Bateson, W. 1909. *Mendel's Principles of Heredity*. Cambridge University Press, Cambridge.

Batterham, P., A. G. Davies, A. Y. Game, and J. A. McKenzie. 1996. Asymmetry—where evolutionary and developmental genetics meet. *Bioessays* 18: 841–845.

Bentley, P. J. 1998. *Comparative Vertebrate Encocrinology*. Cambridge University Press, Cambridge.

Bhattacharyya, M. K., A. M. Smith, T. H. Ellis, C. Hedley, and C. Martin. 1990. The wrinkled-seed character of pea described by Mendel is caused by a transposon-like insertion in a gene encoding starch-branching enzyme. *Cell* 60:115–122.

Bienz, M., G. Saari, G. Tremml, J. Muller, B. Zust, and P. A. Lawrence. 1988. Differential regulation of *Ultrabithorax* in two germ layers of *Drosophila*. *Cell* 53:567–576.

Blomme, T., K. Vandepoele, S. De Bodt, C. Simillion, S. Maere, and Y. Van de Peer. 2006. The gain and loss of genes during 600 million years of vertebrate evolution. *Genome Biology* 7:R43.

Bonner, J. T. 1958. *The Evolution of Development*. Cambridge University Press, New York.

Bonner, J. T. 1982. *Evolution and Development*. Springer-Verlag, Berlin.

Borman, A. M., S. Paulous, and F. Clavel. 1996. Resistance of human immunodeficiency virus type 1 to protease inhibitors: selection of resistance mutations in the presence and absence of the drug. *Journal of General Virology* 77 (Pt. 3):419–426.

Boucher, C. A., E. O'Sullivan, J. W. Mulder, C. Ramautarsing, P. Kellam, G. Darby, J. M. Lange, J. Goudsmit, and B. A. Larder. 1992. Ordered appearance of zidovudine resistance mutations during treatment of 18 human immunodeficiency virus-positive subjects. *The Journal of Infectious Disease* 165:105–110.

Box, N. F., J. R. Wyeth, L. E. O'Gorman, N. G. Martin, and R. A. Sturm. 1997. Characterization of melanocyte stimulating hormone receptor variant alleles in twins with red hair. *Human Molecular Genetics* 6:1891–1897.

Brodie, III, E. D. 1992. Correlational selection for color pattern and antipredator behavior in the garter snake *Thamnophis ordinoides*. *Evolution* 46:1284–1298.

Burch, C. L., and L. Chao. 1999. Evolution by small steps and rugged landscapes in the RNA virus $\phi6$. *Genetics* 151:921–927.

Bustamante, C. D., A. Fledel-Alon, S. Williamson, R. Nielsen, M. T. Hubisz, S. Glanowski, D. M. Tanenbaum, T. J. White, J. J. Sninsky, R. D. Hernandez, D. Civello, M. D. Adams, M. Cargill, and A. G. Clark. 2005. Natural selection on protein-coding genes in the human genome. *Nature* 437:1153–1157.

Cai, J., R. Zhao, H. Jiang, and W. Wang. 2008. *De novo* origination of a new protein-coding gene in *Saccharomyces cerevisiae*. *Genetics* 179:487–496.

Carbone, M. A., A. Llopart, M. deAngelis, J. A. Coyne, and T. F. Mackay. 2005. Quantitative trait loci affecting the difference in pigmentation between *Drosophila yakuba* and *D. santomea*. *Genetics* 171:211–225.

228

Carroll, S. B. 2003. Genetics and the making of *Homo sapiens. Nature* 422:849–857.

Carroll, S. B. 2008. Evo-devo and an expanding evolutionary synthesis: a genetic theory of morphological evolution. *Cell* 134:25–36.

Carroll, S. B., J. K. Grenier, and S. D. Weatherbee. 2001. *From DNA to Diversity: Molecular Genetics and the Evolution of Animal Design, 2nd edition.* Blackwell Publishing, Malden, MA.

Carson, H. L., and A. R. Templeton. 1984. Genetic revolution in relation to speciation phenomena: The founding of new populations. *Annual Review of Ecology and Systematics* 15:97–131.

Casanova, J., E. Sanchez-Herrero, and G. Morata. 1985. Prothoracic transformation and functional structure of the *Ultrabithorax* gene of *Drosophila. Cell* 42:663–669.

Chanut-Delalande, H., I. Fernandes, F. Roch, F. Payre, and S. Plaza. 2006. *Shavenbaby* couples patterning to epidermal cell shape control. *PLoS Biology* 4:e290.

Charlesworth, B. 1992. Evolutionary rates in partially self-fertilizing species. *The American Naturalist* 140:126–148.

Chen, L., A. L. DeVries, and C. H. Cheng. 1997. Evolution of antifreeze glycoprotein gene from a *trypsinogen* gene in Antarctic notothenioid fish. *Proceedings of the National Academy of Sciences of the United States of America* 94:3811–3816.

Cheverud, J. M., and E. J. Routman. 1995. Epistasis and its contribution to genetic variance components. *Genetics* 139:1455–1461.

Chevillon, C., D. Bourguet, F. Rousset, N. Pasteur, and M. Raymond. 1997. Pleiotropy of adaptive changes in populations: comparisons among insecticide resistance genes in *Culex pipiens. Genetics Research* 70:195–203.

Chippindale, A. K. 2006. Experimental evolution. Pp. 482-501 *in* C. W. Fox and J. B. Wolf, eds. *Evolutionary Genetics: Concepts and Case Studies.* Oxford University Press, Oxford.

Chung, H., M. R. Bogwitz, C. McCart, A. Andrianopoulos, R. H. Ffrench-Constant, P. Batterham, and P. J. Daborn. 2007. *Cis*-regulatory elements in the *Accord* retrotransposon result in tissue-specific expression of the *Drosophila melanogaster* insecticide resistance gene *Cyp6g1. Genetics* 175:1071–1077.

Clark, R. M., T. N. Wagler, P. Quijada, and J. Doebley. 2006. A distant upstream enhancer at the maize domestication gene *tb1* has pleiotropic effects on plant and inflorescent architecture. *Nature Genetics* 38:594–597.

Cohn, M. J., and C. Tickle. 1999. Developmental basis of limblessness and axial patterning in snakes. *Nature* 399:474–479.

Colosimo, P. F., K. E. Hosemann, S. Balabhadra, G. Villarreal, Jr., M. Dickson, J. Grimwood, J. Schmutz, R. M. Myers, D. Schluter, and D. M. Kingsley. 2005. Widespread parallel evolution in sticklebacks by repeated fixation of *Ectodysplasin* alleles. *Science* 307:1928–1933.

Cook, L. M. 2003. The rise and fall of the *carbonaria* form of the peppered moth. *The Quarterly Review of Biology* 78:399–417.

Cook, L. M., G. S. Mani, and M. E. Varley. 1986. Postindustrial melanism in the peppered moth. *Science* 231:611–613.

Cooper, T. F., E. A. Ostrowski, and M. Travisano. 2007. A negative relationship between mutation pleiotropy and fitness effect in yeast. *Evolution* 61:1495–1499.

Coyne, J. A., N. H. Barton, and M. Turelli. 2000. Is Wright's shifting balance process important in evolution? *Evolution* 54:306–317.

Coyne, J. A., and B. Charlesworth. 1997. Genetics of a pheromonal difference affecting sexual isolation between *Drosophila mauritiana* and *D. sechellia*. *Genetics* 145:1015–1030.

Coyne, J. A., and H. A. Orr. 2004. *Speciation*. Sinauer Associates, Inc., Sunderland, MA.

Crow, J. E., and M. Kimura. 1970. *An Introduction to Population Genetics Theory*. Harper & Row Publishers, New York.

Crow, J. F. 1956. Genetics of DDT resistance in *Drosophila*. International Genetics Symposia. The Organizing Committee, International Genetics Symposia, Tokyo, Japan.

Crow, J. F., and M. J. Simmons. 1983. The mutation load in *Drosophila*. Pp. 1–35 *in* M. Ashburner, H. L. Carson, and J. N. Thompson, Jr., eds. *The Genetics and Biology of Drosophila*. Academic Press, New York.

da Silva, L. B., D. F. Leite, V. L. Valente, and C. Rohde. 2005. Mating activity of *yellow* and *sepia Drosophila willistoni* mutants. *Behavioural Processes* 70:149–155.

Darwin, C. 1859. *The Origin of Species by Means of Natural Selection or the Preservation of Favored Races in the Struggle for Life*. The Modern Library, New York.

Davidson, E. H. 2001. *Genomic Regulatory Systems*. Academic Press, San Diego.

Davidson, E. H. 2006. *The Regulatory Genome: Gene Regulatory Networks in Development and Evolution*. Academic Press, Burlington, VT.

Davies, A. G., A. Y. Game, Z. Chen, T. J. Williams, S. Goodall, J. L. Yen, J. A. McKenzie, and P. Batterham. 1996. Scalloped wings is the *Lucilia cuprina Notch* homologue and a candidate for the modifier of fitness and asymmetry of diazinon resistance. *Genetics* 143:1321–1337.

Davis, G. K., D. G. Srinivasan, P. J. Wittkopp, and D. L. Stern. 2007. The function and regulation of *Ultrabithorax* in the legs of *Drosophila melanogaster*. *Developmental Biology* 308:621–631.

De Beer, G. 1940. *Embryos and Ancestors*. Oxford University Press, Oxford.

de Visser, J. A., and D. E. Rozen. 2006. Clonal interference and the periodic selection of new beneficial mutations in *Escherichia coli*. *Genetics* 172:2093–2100.

Demuth, J. P., T. De Bie, J. E. Stajich, N. Cristianini, and M. W. Hahn. 2006. The evolution of mammalian gene families. *PLoS One* 1:e85.

de Vicente, M. C., and S. D. Tanksley. 1993. QTL analysis of transgressive segregation in an interspecific tomato cross. *Genetics* 134:585–596.

DeWitt, T. J., and S. M. Scheiner. 2004. *Phenotypic Plasticity: Functional and Conceptual Approaches*. Oxford University Press, Oxford.

Dickinson, W. J., Y. Tang, K. Schuske, and M. Akam. 1993. Conservation of molecular prepatterns during the evolution of cuticle morphology in *Drosophila* larvae. *Evolution* 47:1396–1406.

Dobzhansky, T. 1927. Studies on the manifold effect of certain genes in *Drosophila melanogaster. Zeitschrift für induktive Abstammungsund Vererbungslehre* 43:330–388.

Dobzhansky, T. 1973. Nothing in biology makes sense except in the light of evolution. *The American Biology Teacher* 35:125–129.

Dobzhansky, T., and A. M. Holz. 1943. A re-examination of the problem of manifold effects of genes in *Drosophila melanogaster. Genetics* 28:295–303.

Doebley, J., A. Stec, and C. Gustus. 1995. *teosinte branched1* and the origin of maize: Evidence for epistasis and the evolution of dominance. *Genetics* 141:333–346.

Doebley, J., A. Stec, and L. Hubbard. 1997. The evolution of apical dominance in maize. *Nature* 386:485–488.

Drapeau, M. D., S. A. Cyran, M. M. Viering, P. K. Geyer, and A. D. Long. 2006. A *cis*-regulatory sequence within the *yellow* locus of *Drosophila melanogaster* required for normal male mating success. *Genetics* 172:1009–1030.

Dunham, M. J., H. Badrane, T. Ferea, J. Adams, P. O. Brown, F. Rosenzweig, and D. Botstein. 2002. Characteristic genome rearrangements in experimental evolution of *Saccharomyces cerevisiae. Proceedings of the National Academy of Sciences of the United States of America* 99:16144–16149.

Dykhuizen, D., and D. L. Hartl. 1980. Selective neutrality of 6PGD allozymes in *E. coli* and the effects of genetic background. *Genetics* 96:801–817.

Eldredge, N., and S. J. Gould. 1972. Punctuated equilibria: An alternative to phyletic gradualism. Pp. 82–115. *in* T. J. M. Schopf, ed. *Models in Paleobiology.* Freeman, Cooper & Co., San Francisco.

Elena, S. F., and R. E. Lenski. 2003. Evolution experiments with microorganisms: the dynamics and genetic bases of adaptation. *Nature Reviews Genetics* 4:457–469.

Endler, J. A. 1986. *Natural Selection in the Wild.* Princeton University Press, Princeton, NJ.

Ewens, W. J. 1967. The probability of survival of a new mutant in a fluctuating environment. *Heredity* 22:438–443.

Falconer, D. S., and T. F. C. Mackay. 1996. *Introduction to Quantitative Genetics.* Addison Wesley Longman Limited, Harlow.

Felsenstein, J. 1974. The evolutionary advantage of recombination. *Genetics* 78:737–756.

Fisher, R. A. 1918. The correlation between relatives on the supposition of Mendelian inheritance. *Transactions of the Royal Society of Edinburgh* 52:399–433.

Fisher, R. A. 1930. *The Genetical Theory of Natural Selection.* Oxford University Press, Oxford.

Fisher, R. A. 1958. *The Genetical Theory of Natural Selection.* Dover, New York.

Force, A., M. Lynch, F. B. Pickett, A. Amores, Y. L. Yan, and J. Postlethwait. 1999. Preservation of duplicate genes by complementary, degenerative mutations. *Genetics* 151:1531–1545.

Frandberg, P. A., M. Doufexis, S. Kapas, and V. Chhajlani. 1998. Human pigmentation phenotype: a point mutation generates nonfunctional MSH receptor. *Biochemical and Biophysical Research Communications* 245:490–492.

Frankham, R. 1995. Effective population size/adult population size ratios in wildlife: a review. *Genetical Research* 66:95–107.

Gadau, J., R. E. Page, and J. H. Werren. 2002. The genetic basis of the interspecific differences in wing size in *Nasonia* (Hymenoptera; Pteromalidae): major quantitative trait loci and epistasis. *Genetics* 161:673–684.

Galant, R., and S. B. Carroll. 2002. Evolution of a transcriptional repression domain in an insect *Hox* protein. *Nature* 415:910–913.

Galant, R., C. M. Walsh, and S. B. Carroll. 2002. *Hox* repression of a target gene: *extradenticle*-independent, additive action through multiple monomer binding sites. *Development* 129:3115–3126.

Gazzani, S., A. R. Gendall, C. Lister, and C. Dean. 2003. Analysis of the molecular basis of flowering time variation in *Arabidopsis* accessions. *Plant Physiology* 132:1107–1114.

Gebelein, B., J. Culi, H. D. Ryoo, W. Zhang, and R. S. Mann. 2002. Specificity of *Distalless* repression and limb primordia development by abdominal Hox proteins. *Developmental Cell* 3:487–498.

Gerhart, J., and M. Kirschner. 1997. *Cells, Embryos, and Evolution.* Blackwell Science, Malden.

Gerke, J., K. Lorenz, and B. Cohen. 2009. Genetic interactions between transcription factors cause natural variation in yeast. *Science* 323:498–501.

Gerke, J. P., C. T. Chen, and B. A. Cohen. 2006. Natural isolates of *Saccharomyces cerevisiae* display complex genetic variation in sporulation efficiency. *Genetics* 174:985–987.

Gerrish, P. J., and R. E. Lenski. 1998. The fate of competing beneficial mutations in an asexual population. *Genetica* 102/103:127–144.

Geyer, P. K., and V. G. Corces. 1987. Separate regulatory elements are responsible for the complex pattern of tissue-specific and developmental transcription of the *yellow* locus in *Drosophila melanogaster*. *Genes & Development* 1:996–1004.

Geyer, P. K., and V. G. Corces. 1992. DNA position-specific repression of transcription by a *Drosophila* zinc finger protein. *Genes & Development* 6:1865–1873.

Gibson, G., and S. van Helden. 1997. Is function of the *Drosophila* homeotic gene *Ultrabithorax* canalized? *Genetics* 147:1155–1168.

Gibson, G., M. Wemple, and S. van Helden. 1999. Potential variance affecting homeotic *Ultrabithorax* and *Antennapedia* phenotypes in *Drosophila melanogaster*. *Genetics* 151:1081–1091.

Gillespie, J. H. 1973. Natural selection with varying selection coefficients—a haploid model. *Genetical Research Cambridge* 21:115–120.

Gillespie, J. 1974a. Polymorphism in patchy environments. *The American Naturalist* 108:145–151.

Gillespie, J. 1974b. The role of environmental grain in the maintenance of genetic variation. *The American Naturalist* 108:831–836.

Gillespie, J. H. 1974c. Natural selection for within-generation variance in offspring number. *Genetics* 76:601–606.

Gillespie, J. H. 1975. Natural selection for within-generation variance in offspring number II. Discrete haploid models. *Genetics* 81:403–413.

Gillespie, J. H. 1977. Natural selection for variances in offspring numbers: A new evolutionary principle. *The American Naturalist* 111:1010–1014.

Gillespie, J. H. 1991. *The Causes of Molecular Evolution*. Oxford University Press, New York.

Gillespie, J. H. 2004. *Population Genetics: A Concise Guide*. The Johns Hopkins University Press, Baltimore.

Gillespie, J. H. 2006. Stochastic processes in evolution. Pp. 65–79 *in* C. W. Fox and J. B. Wolf, eds. *Evolutionary Genetics: Concepts and Case Studies*. Oxford University Press, New York.

Gillespie, J., and C. Langley. 1976. Multilocus behavior in random environments. I. Random Levene models. *Genetics* 82:123–137.

Goldschmidt, R. 1940. *The Material Basis of Evolution*. Yale University Press, New Haven.

Gompel, N., B. Prud'homme, P. J. Wittkopp, V. A. Kassner, and S. B. Carroll. 2005. Chance caught on the wing: *cis*-regulatory evolution and the origin of pigment patterns in *Drosophila*. *Nature* 433:481–487.

Goodnight, C. J. 1987. On the effect of founder events on the epistatic genetic variance. *Evolution* 41:80–91.

Goodwin, B. C., N. Holder, and C. C. Wylie, eds. 1983. *Development and Evolution. (British Society for Developmental Biology Symposia)*. Cambridge University Press, Cambridge.

Gould, S. J., and N. Eldredge. 1977. Punctuated equilibria: the tempo and mode of evolution reconsidered. *Paleobiology* 3:115–151.

Gould, S. J., and N. Eldredge. 1986. Punctuated equilibrium at the third stage. *Systematic Zoology* 35:143–148.

Granot, D., J. P. Margolskee, and G. Simchen. 1989. A long region upstream of the *IME1* gene regulates meiosis in yeast. *Molecular and General Genetics* 218:308–314.

Grant, B. R., and P. R. Grant. 1993. Evolution of Darwin's finches caused by a rare climatic event. *Proceedings of the Royal Society B: Biological Sciences* 251:111–117.

Grant, B. S. 2004. Allelic melanism in American and British peppered moths. *Journal of Heredity* 95:97–102.

Grant, P. R., and B. R. Grant. 2002. Unpredictable evolution in a 30-year study of Darwin's finches. *Science* 296:707–711.

Grant, P. R., and B. R. Grant. 2008. *How and Why Species Multiply: The Radiation of Darwin's Finches*. Princeton University Press, Princeton.

Griswold, C. K., and M. C. Whitlock. 2003. The genetics of adaptation: the roles of pleiotropy, stabilizing selection and drift in shaping the distribution of bidirectional fixed mutational effects. *Genetics* 165:2181–2192.

Grobet, L., L. J. Martin, D. Poncelet, D. Pirottin, B. Brouwers, J. Riquet, A. Schoeberlein, S. Dunner, F. Menissier, J. Massabanda, R. Fries, R. Hanset, and M. Georges. 1997. A deletion in the bovine *myostatin* gene causes the double-muscled phenotype in cattle. *Nature Genetics* 17:71–74.

Grobet, L., D. Poncelet, L. J. Royo, B. Brouwers, D. Pirottin, C. Michaux, F. Menissier, M. Zanotti, S. Dunner, and M. Georges. 1998. Molecular definition of an allelic series of mutations disrupting the *myostatin* function and causing double-muscling in cattle. *Mammalian Genome* 9:210–213.

Haldane, J. B. S. 1924. A mathematical theory of natural and artificial selection, Part I. *Transactions of the Cambridge Philosophical Society* 23:19–41.

Haldane, J. B. S. 1927. A mathematical theory of natural and artificial selection, part V: Selection and mutation. *Proceedings of the Cambridge Philosophical Society* 28:838–844.

Haldane, J. B. S. 1932. *The Causes of Evolution*. Princeton University Press edition (1990), Princeton.

Haldane, J. B. S. 1937. The effect of variation on fitness. *The American Naturalist* 71:337–349.

Harding, R. M., E. Healy, A. J. Ray, N. S. Ellis, N. Flanagan, C. Todd, C. Dixon, A. Sajantila, I. J. Jackson, M. A. Birch-Machin, and J. L. Rees. 2000. Evidence for variable selective pressures at *MC1R*. *The American Journal of Human Genetics* 66:1351–1361.

Hartl, D. L., and A. G. Clark. 1997. *Principles of Population Genetics*. Sinauer Press, Sunderland, MA.

Hastings, A. 1989. Linkage disequilibrium and genetic variances under mutation-selection balance. *Genetics* 121:857– 860.

Hatini, V., and S. DiNardo. 2001. Divide and conquer: Pattern formation in *Drosophila* embryonic epidermis. *Trends in Genetics* 17:574–579.

Haubst, N., J. Favor and M. Götz. 2006. The role of *Pax6* in the nervous system during development and in adulthood: Master control regulator or modular function? *In* Gerald Thiel, ed. *Transcription Factors in the Nervous System*. Wiley-VCH Verlag GmbH & Co. KGaA, Weinheim, Germany.

Hermisson, J., and P. S. Pennings. 2005. Soft sweeps: molecular population genetics of adaptation from standing genetic variation. *Genetics* 169:2335–2352.

Hersh, B. M., and S. B. Carroll. 2005. Direct regulation of *knot* gene expression by Ultrabithorax and the evolution of *cis*-regulatory elements in *Drosophila*. *Development* 132:1567–1577.

Hewitt, G. 2000. The genetic legacy of the Quaternary ice ages. *Nature* 405: 907–913.

Hill, W. G., and A. Robertson. 1966. The effect of linkage on limits to artificial selection. *Genetics Research* 8:269–294.

Hittinger, C. T., D. L. Stern, and S. B. Carroll. 2005. Pleiotropic functions of a conserved insect-specific Hox peptide motif. *Development* 132:5261–5270.

Hoekstra, H. E., and J. A. Coyne. 2007. The locus of evolution: evo devo and the genetics of adaptation. *Evolution* 61:995–1016.

Hoekstra, H. E., J. M. Hoekstra, D. Berrigan, S. N. Vignieri, A. Hoang, C. E. Hill, P. Beerli, and J. G. Kingsolver. 2001. Strength and tempo of directional selection in the wild. *Proceedings of the National Academy of Sciences of the United States of America* 98:9157–9160.

Högland, J., and R. V. Alatalo. 1995. *Leks*. Princeton University Press, Princeton.

Holland, P. W. H. 1999. Gene duplication: Past, present and future. *Seminars in Cell and Developmental Biology* 10:541–547.

Huerta-Sanchez, E., R. Durrett, and C. D. Bustamante. 2008. Population genetics of polymorphism and divergence under fluctuating selection. *Genetics* 178:325–337.

Hurst, L. D., and N. G. C. Smith. 1999. Do essential genes evolve slowly? *Current Biology* 9:747–750.

Huxley, J. S. 1932. *Problems of Relative Growth*. Methuen & Co., London.

Imaizumi, T., and S. A. Kay. 2006. Photoperiodic control of flowering: not only by coincidence. *Trends in Plant Science* 11:550–558.

Immergluck, K., P. A. Lawrence, and M. Bienz. 1990. Induction across germ layers in *Drosophila* mediated by a genetic cascade. *Cell* 62:261–268.

Ingram, C. J., M. F. Elamin, C. A. Mulcare, M. E. Weale, A. Tarekegn, T. O. Raga, E. Bekele, F. M. Elamin, M. G. Thomas, N. Bradman, and D. M. Swallow. 2007. A novel polymorphism associated with lactose tolerance in Africa: multiple causes for lactase persistence? *Human Genetics* 120:779–788.

Innan, H., and Y. Kim. 2004. Pattern of polymorphism after strong artificial selection in a domestication event. *Proceedings of the National Academy of Sciences of the United States of America* 101:10667–10672.

Jaillon, O., J. M. Aury, F. Brunet, J. L. Petit, N. Stange-Thomann, E. Mauceli, L. Bouneau, C. Fischer, C. Ozouf-Costaz, A. Bernot, S. Nicaud, D. Jaffe, S. Fisher, G. Lutfalla, C. Dossat, B. Segurens, C. Dasilva, M. Salanoubat, M. Levy, N. Boudet, S. Castellano, V. Anthouard, C. Jubin, V. Castelli, M. Katinka, B. Vacherie, C. Biemont, Z. Skalli, L. Cattolico, J. Poulain, V. De Berardinis, C. Cruaud, S. Duprat, P. Brottier, J. P. Coutanceau, J. Gouzy, G. Parra, G. Lardier, C. Chapple, K. J. McKernan, P. McEwan, S. Bosak, M. Kellis, J. N. Volff, R. Guigo, M. C. Zody, J. Mesirov, K. Lindblad-Toh, B. Birren, C. Nusbaum, D. Kahn, M. Robinson-Rechavi, V. Laudet, V. Schachter, F. Quetier, W. Saurin, C. Scarpelli, P. Wincker, E. S. Lander, J. Weissenbach, and H. Roest Crollius. 2004. Genome duplication in the teleost fish *Tetraodon nigroviridis* reveals the early vertebrate proto-karyotype. *Nature* 431:946–957.

Jeong, S., M. Rebeiz, P. Andolfatto, T. Werner, J. True, and S. B. Carroll. 2008. The evolution of gene regulation underlies a morphological difference between two *Drosophila* sister species. *Cell* 132:783–793.

Jia, L., M. T. Clegg, and T. Jiang. 2003. Excess non-synonymous substitutions suggest that positive selection episodes occurred during the evolution of

DNA-binding domains in the *Arabidopsis R2R3-MYB* gene family. *Plant Molecular Biology* 52:627–642.

Jia, L., M. T. Clegg, and T. Jiang. 2004. Evolutionary dynamics of the DNA-binding domains in putative *R2R3-MYB* genes identified from rice subspecies *indica* and *japonica* genomes. *Plant Physiology* 134:575–585.

Jiang, N., Z. Bao, X. Zhang, S. R. Eddy, and S. R. Wessler. 2004. Pack-MULE transposable elements mediate gene evolution in plants. *Nature* 431:569–573.

Johanson, U., J. West, C. Lister, S. Michaels, R. Amasino, and C. Dean. 2000. Molecular analysis of *FRIGIDA*, a major determinant of natural variation in *Arabidopsis* flowering time. *Science* 290:344–347.

John, P. R., K. Makova, W. H. Li, T. Jenkins, and M. Ramsay. 2003. DNA polymorphism and selection at the *Melanocortin-1 Receptor* gene in normally pigmented southern African individuals. *Annals of the New York Academy of Science* 994:299–306.

Johnson, A. D., D. Fitzsimmons, J. Hagman, and H. M. Chamberlin. 2001. EGL-38 Pax regulates the *ovo*-related gene *lin-48* during *Caenorhabditis elegans* organ development. *Development* 128:2857–2865.

Joly, D., C. Bazin, L. W. Zeng, and R. S. Singh. 1997. Genetic basis of sperm and testis length differences and epistatic effect on hybrid inviability and sperm motility between *Drosophila simulans* and *D. sechellia*. *Heredity* 78:354–362.

Jones, C. D., and D. J. Begun. 2005. Parallel evolution of chimeric fusion genes. *Proceedings of the National Academy of Sciences of the United States of America* 102:11373–11378.

Jung, J. H., Y. H. Seo, P. J. Seo, J. L. Reyes, J. Yun, N. H. Chua, and C. M. Park. 2007. The GIGANTEA-regulated microRNA172 mediates photoperiodic flowering independent of CONSTANS in *Arabidopsis*. *The Plant Cell* 19:2736–2748.

Kacser, H., and J. A. Burns. 1981. The molecular basis of dominance. *Genetics* 97:639–666.

Kambadur, R., M. Sharma, T. P. Smith, and J. J. Bass. 1997. Mutations in *myostatin* (*GDF8*) in double-muscled Belgian Blue and Piedmontese cattle. *Genome Research* 7:910–916.

Kaneshiro, K. Y. 1988. Speciation in the Hawaiian *Drosophila*. *BioScience* 38:258–263.

Kao, K. C., and G. Sherlock. 2008. Molecular characterization of clonal interference during adaptive evolution in asexual populations of *Saccharomyces cerevisiae*. *Nature Genetics* 40:1499–1504.

Kavanagh, K. D., A. R. Evans, and J. Jernvall. 2007. Predicting evolutionary patterns of mammalian teeth from development. *Nature* 449:427–432.

Kellam, P., C. A. Boucher, J. M. Tijnagel, and B. A. Larder. 1994. Zidovudine treatment results in the selection of human immunodeficiency virus type 1 variants whose genotypes confer increasing levels of drug resistance. *Journal of General Virology* 75:341–351.

Kellis, M., B. W. Birren, and E. S. Lander. 2004. Proof and evolutionary analysis of ancient genome duplication in the yeast *Saccharomyces cerevisiae*. *Nature* 428:617–624.

Kelsh, R., R. O. Weinzierl, R. A. White, and M. Akam. 1994. Homeotic gene expression in the locust *Schistocerca*: an antibody that detects conserved epitopes in Ultrabithorax and abdominal-A proteins. *Developmental Genetics* 15:19–31.

Kerridge, S., and G. Morata. 1982. Developmental effects of some newly induced *Ultrabithorax* alleles of *Drosophila*. *Journal of Embryology and Experimental Morphology* 68:211–234.

Khila, A., A. El Haidani, A. Vincent, F. Payre, and S. I. Souda. 2003. The dual function of *ovo/shavenbaby* in germline and epidermis differentiation is conserved between *Drosophila melanogaster* and the olive fruit fly *Bactrocera oleae*. *Insect Biochemisty and Molecular Biology* 33:691–699.

Kim, Y., and H. A. Orr. 2005. Adaptation in sexuals vs. asexuals: clonal interference and the Fisher-Muller model. *Genetics* 171:1377–1386.

Kimura, M. 1962. On the probability of fixation of mutant genes in a population. *Genetics* 47:713–719.

Kimura, M. 1983. *The Neutral Theory of Molecular Evolution*. Cambridge University Press, Cambridge.

Kimura, M., and T. Ohta. 1969. The average number of generations until fixation of a mutant gene in a finite population. *Genetics* 61:763–771.

Kingsolver, J. G., H. E. Hoekstra, J. M. Hoekstra, D. Berrigan, S. N. Wignieri, C. E. Hill, A. Hoang, P. Gibert, and P. Beerli. 2001. The strength of phenotypic selection in natural populations. *The American Naturalist* 157:245–261.

Knowles, D. G., and A. McLysaght. 2009. Recent *de novo* origin of human protein-coding genes. *Genome Research* 19:1752–1759.

Kobayashi, S., N. Goto-Yamamoto, and H. Hirochika. 2004. Retrotransposon-induced mutations in grape skin color. *Science* 304:982.

Kondrashov, F. A., and E. V. Koonin. 2004. A common framework for understanding the origin of genetic dominance and evolutionary fates of gene duplications. *Trends in Genetics* 20:287–290.

Kopp, M., and J. Hermisson. 2009. The genetic basis of phenotypic adaptation II: The distribution of adaptive substitutions in the moving optimum model. *Genetics* 183:1453–1476.

Korpimäki, E., and C. J. Krebs. 1996. Predation and population cycles of small mammals. *BioScience* 46:754–764.

Korves, T. M., K. J. Schmid, A. L. Caicedo, C. Mays, J. R. Stinchcombe, M. D. Purugganan, and J. Schmitt. 2007. Fitness effects associated with the major flowering time gene *FRIGIDA* in *Arabidopsis thaliana* in the field. *The American Naturalist* 169:E141–157.

Krebs, C. J., R. Boonstra, S. Boutin, and A. R. E. Sinclair. 2001. What drives the 10-year cycle of Snowshoe Hares? *BioScience* 51:25–35.

Kroymann, J., and T. Mitchell-Olds. 2005. Epistasis and balanced polymorphism influencing complex trait variation. *Nature* 435:95–98.

Kruger, O., J. Lindstrom, and W. Amos. 2001. Maladaptive mate choice maintained by heterozygote advantage. *Evolution* 55:1207–1214.

Kruglyak, L., and D. A. Nickerson. 2001. Variation is the spice of life. *Nature Genetics* 27:234–236.

Kuittinen, H., A. Niittyvuopio, P. Rinne, and O. Savolainen. 2008. Natural variation in *Arabidopsis lyrata* vernalization requirement conferred by a *FRIGIDA* indel polymorphism. *Molecular Biology and Evolution* 25:319–329.

Lande, R. 1975. The maintenance of genetic variability by mutation in a polygenic character with linked loci. *Genetics Research* 26:221–235.

Lawrence, P. A. 1992. *The Making of a Fly*. Blackwell Scientific Publications, Oxford.

Le Corre, V., F. Roux, and X. Reboud. 2002. DNA polymorphism at the *FRIGIDA* gene in *Arabidopsis thaliana*: extensive nonsynonymous variation is consistent with local selection for flowering time. *Molecular Biology and Evolution* 19:1261–1271.

Lees, D. R. 1968. Genetic control of the melanic form *insularia* of the peppered moth *Biston betularia* (L.). *Nature* 220:1249–1250.

Lemons, D., and W. McGinnis. 2006. Genomic evolution of *Hox* gene clusters. *Science* 313:1918–1922.

Lenski, R. E., and M. Travisano. 1994. Dynamics of adaptation and diversification: a 10,000-generation experiment with bacterial populations. *Proceedings of the National Academy of Sciences of the United States of America* 91:6808–6814.

Levy, Y. Y., and C. Dean. 1998. The transition to flowering. *The Plant Cell* 10:1973–1990.

Lewis, E. B. 1978. A gene complex controlling segmentation in *Drosophila*. *Nature* 276:565–570.

Li, W.-H. 1997. *Molecular Evolution*. Sinauer Associates, Sunderland, MA.

Liu, J., J. M. Nercer, L. F. Stam, G. C. Gibson, Z.-B. Zeng, and C. C. Laurie. 1996. Genetic analysis of a morphological shape difference in the male genitalia of *Drosophila simulans* and *D. mauritiana*. *Genetics* 142:1129–1145.

Long, A. D., R. F. Lyman, A. H. Morgan, C. H. Langley, and T. F. C. Mackay. 2000. Both naturally occurring insertions of transposable elements and intermediate frequency polymorphisms at the *achaete-scute* complex are associated with variation in bristle number in *Drosophila melanogaster*. *Genetics* 154:1255–1269.

Long, A. D., S. L. Mullaney, L. A. Reid, J. D. Fry, C. H. Langley, and T. F. C. Mackay. 1995. High resolution mapping of genetic factors affecting abdominal bristle number in *Drosophila melanogaster*. *Genetics* 139:1273–1291.

Long, M., and C. H. Langley. 1993. Natural selection and the origin of *jingwei*, a chimeric processed functional gene in *Drosophila*. *Science* 260:91–95.

Longabaugh, W. J., E. H. Davidson, and H. Bolouri. 2005. Computational representation of developmental genetic regulatory networks. *Developmental Biology* 283:1–16.

Longabaugh, W. J., E. H. Davidson, and H. Bolouri. 2009. Visualization, documentation, analysis, and communication of large-scale gene regulatory networks. *Biochimica et Biophysiea Acta* 1789:363–374.

Lunzer, M., S. P. Miller, R. Felsheim, and A. M. Dean. 2005. The biochemical architecture of an ancient adaptive landscape. *Science* 310:499–501.

Lynch, M., and J. S. Conery. 2000. The evolutionary fate and consequences of duplicate genes. *Science* 290:1151–1155.

Lynch, M., and A. Force. 2000. The probability of duplicate gene preservation by subfunctionalization. *Genetics* 154:459–473.

Lynch, M., and B. Walsh. 1998. *Genetics and Analysis of Quantitative Traits.* Sinauer Associates, Inc., Sunderland, MA.

Lynch, V. J., and G. P. Wagner. 2008. Resurrecting the role of transcription factor change in developmental evolution. *Evolution* 62:2131–2154.

Mackay, T. F. 2004. The genetic architecture of quantitative traits: lessons from *Drosophila. Current Opinion in Genetics and Development* 14:253–257.

Mackay, T. F., and C. H. Langley. 1990. Molecular and phenotypic variation in the *achaete-scute* region of *Drosophila melanogaster. Nature* 348:64–66.

Mackay, T. F. C. 2001. Quantitative trait loci in *Drosophila. Nature Reviews Genetics* 2:11–20.

Mackenzie, A., J. D. Reynolds, V. J. Brown, and W. J. Sutherland. 1995. Variation in male mating success on leks. *The American Naturalist* 145:633–652.

MacLuliich, D. A. 1937. Fluctuations in the numbers of the Varying Hare (*Lepus americanus*). *University of Toronto Studies: Biological Series* 43:1–136.

Mahfooz, N., N. Turchyn, M. Mihajlovic, S. Hrycaj, and A. Popadic. 2007. Ubx regulates differential enlargement and diversification of insect hind legs. *PLoS ONE* 2:e866.

Malmberg, R. L., and R. Mauricio. 2005. QTL-based evidence for the role of epistasis in evolution. *Genetics Research* 86:89–95.

Maynard Smith, J., R. Burian, S. Kauffman, P. Alberch, J. Cambell, B. Goodwin, R. Lande, D. Raup, and L. Wolpert. 1985. Developmental constraints and evolution. *The Quarterly Review of Biology* 60:265–287.

Mayr, E. 1954. Change of genetic environment and evolution. Pp. 157–180 *in* J. Huxley, A. C. Hardy, and E. B. Ford, eds. *Evolution as a Process.* Allen and Unwin, London.

Mayr, E. 1961. Cause and effect in biology. *Science* 134:1501–1506.

Mayr, E. 1963. *Animal Species and Evolution.* The Belknap Press of Harvard University, Cambridge, MA.

McGregor, A. P., V. Orgogozo, I. Delon, J. Zanet, D. G. Srinivasan, F. Payre, and D. L. Stern. 2007. Morphological evolution through multiple *cis*-regulatory mutations at a single gene. *Nature* 448:587–590.

McKenzie, J. A. 1990. Selection at the *Dieldrin Resistance* locus in overwintering populations of *Lucilia cuprina* (Wiedemann). *Australian Journal of Zoology* 38:493–501.

McKenzie, J. A., and G. M. Clarke. 1988. Diazinon resistance, fluctuating asymmetry and fitness in the australian sheep blowfly, *Lucilia cuprina*. *Genetics* 120:213–220.

McPherron, A. C., and S. J. Lee. 1997. Double muscling in cattle due to mutations in the *myostatin* gene. *Proceedings of the National Academy of Sciences of the United States of America* 94:12457–12461.

Mendel, G. 1865. Versuche über Pflanzen-Hybriden. *Verhandlungen des naturforschenden Vereines in Brünn, Nd. IV für das Jahr 1865* 4:3–47.

Michaels, S. D., and R. M. Amasino. 1999. *FLOWERING LOCUS C* encodes a novel MADS domain protein that acts as a repressor of flowering. *The Plant Cell* 11:949–956.

Michaels, S. D., Y. He, K. C. Scortecci, and R. M. Amasino. 2003. Attenuation of *FLOWERING LOCUS C* activity as a mechanism for the evolution of summer-annual flowering behavior in *Arabidopsis*. *Proceedings of the National Academy of Sciences of the United States of America* 100:10102–10107.

Middleton, R. J., and H. Kacser. 1983. Enzyme variation, metabolic flux and fitness: Alcohol dehydrogenase in *Drosophila melanogaster*. *Genetics* 105:633–650.

Minelli, A. 2003. *The Development of Animal Form: Ontogeny, Morphology, and Evolution*. Cambridge University Press, Cambridge.

Moehring, A. J., A. Llopart, S. Elwyn, J. A. Coyne, and T. F. Mackay. 2006. The genetic basis of prezygotic reproductive isolation between *Drosophila santomea* and *D. yakuba* due to mating preference. *Genetics* 173:215–223.

Mogil, J. S., S. G. Wilson, E. J. Chesler, A. L. Rankin, K. V. Nemmani, W. R. Lariviere, M. K. Groce, M. R. Wallace, L. Kaplan, R. Staud, T. J. Ness, T. L. Glover, M. Stankova, A. Mayorov, V. J. Hruby, J. E. Grisel, and R. B. Fillingim. 2003. The *Melanocortin-1 Receptor* gene mediates female-specific mechanisms of analgesia in mice and humans. *Proceedings of the National Academy of Sciences of the United States of America* 100:4867–4872.

Moore, J. H., and S. M. Williams. 2005. Traversing the conceptual divide between biological and statistical epistasis: systems biology and a more modern synthesis. *Bioessays* 27:637–646.

Morimoto, R. I. 1993. Cells in stress: transcriptional activation of heat shock genes. *Science* 259:1409–1410.

Mosher, D. S., P. Quignon, C. D. Bustamante, N. B. Sutter, C. S. Mellersh, H. G. Parker, and E. A. Ostrander. 2007. A mutation in the *myostatin* gene increases muscle mass and enhances racing performance in heterozygote dogs. *PLoS Genetics* 3:e79.

Muller, H. J. 1922. Variation due to change in the individual gene. *The American Naturalist* 56:32–50.

Muller, H. J. 1932. Some genetic aspects of sex. *The American Naturalist* 66:118–138.

Muller, H. J. 1935. The origination of chromatin deficiencies as minute deletions subject to insertion elsewhere. *Genetica* 17:237–252.

Mustonen, V., and M. Lassig. 2007. Adaptations to fluctuating selection in *Drosophila*. *Proceedings of the National Academy of Sciences of the United States of America* 104:2277–2282.

Mustonen, V., and M. Lassig. 2009. From fitness landscapes to seascapes: nonequilibrium dynamics of selection and adaptation. *Trends in Genetics* 25:111–119.

Nachman, M. W. 2006. Detecting selection at the molecular level. Pp. 103–118 *in* C. W. Fox, and J. B. Wolf, eds. *Evolutionary Genetics: Concepts and Case Studies*. Oxford University Press, Oxford.

Naumann, I. D., ed. 1970. *The Insects of Australia: A Textbook for Students and Research Workers – Volume 2*. Melbourne University Publishing, Carlton, Victoria.

Necker, L. A. 1832. Observations on some remarkable optical phaenomena seen in Switzerland; and on an optical phaenomenon which occurs on viewing a figure of a crystal or geometrical solid. *The London and Edinburgh Philosophical Magazine and Journal of Science* 1:329–337.

Nei, M. 2005. Selectionism and neutralism in molecular evolution. *Molecular Biology and Evolution* 22:2318–2342.

Ng, C. S., A. M. Hamilton, A. Frank, O. Barmina, and A. Kopp. 2008. Genetic basis of sex-specific color pattern variation in *Drosophila malerkotliana*. *Genetics* 180:421–429.

Nijhout, H. F. 1991. *The Development and Evolution of Butterfly Wing Patterns*. Smithsonian Institution Press, Washington, DC.

Nijhout, H. F. 1999. Control mechanisms of polyphenic development. *Bioscience* 49:181–192.

Nijhout, H. F., and S. M. Paulsen. 1997. Developmental models and polygenic characters. *The American Naturalist* 149:394–405.

Noor, M. A. 1999. Reinforcement and other consequences of sympatry. *Heredity* 83 (5):503–508.

Noor, M. A., and J. A. Coyne. 1996. Genetics of a difference in cuticular hydrocarbons between *Drosophila pseudoobscura* and *D. persimilis*. *Genetics Research* 68:117–123.

Nylin, S., and K. Gotthard. 1998. Plasticity in life history traits. *Annual Review of Entomology* 43:63–83.

Odum, E., R. Brewer, and G. W. Barrett. 2005. *Fundamentals of Ecology*. Thomson Brooks/Cole, Belmont, CA.

Ohno, S. 1970. *Evolution by Gene Duplication*. Springer-Verlag, New York.

Orgogozo, V., K. W. Broman, and D. L. Stern. 2006. High-resolution quantitative trait locus mapping reveals sign epistasis controlling ovariole number between two *Drosophila* species. *Genetics* 173:197–205.

Orr, H. A. 1996. Dobzhansky, Bateson, and the genetics of speciation. *Genetics* 144:1331–1335.

Orr, H. A. 1998. The population genetics of adaptation: the distribution of factors fixed during adaptive evolution. *Evolution* 52:935–949.

Orr, H. A. 2000. Adaptation and the cost of complexity. *Evolution* 54:13–20.

Orr, H. A. 2001. The genetics of species differences. *Trends in Ecology and Evolution* 16:343–350.

Orr, H. A., and A. J. Betancourt. 2001. Haldane's sieve and adaptation from the standing genetic variation. *Genetics* 157:875–884.

Ortlund, E. A., J. T. Bridgham, M. R. Redinbo, and J. W. Thornton. 2007. Crystal structure of an ancient protein: evolution by conformational epistasis. *Science* 317:1544–1548.

Otto, S. P. 2004. Two steps forward, one step back: the pleiotropic effects of favoured alleles. *Proceedings of the Royal Society of London Biological Sciences* 271:705–714.

Otto, S. P., and M. C. Whitlock. 1997. The probability of fixation in populations of changing size. *Genetics* 146:723–733.

Overton, P. M., W. Chia, and M. Buescher. 2007. The *Drosophila* HMG-domain proteins SoxNeuro and Dichaete direct trichome formation via the activation of *shavenbaby* and the restriction of Wingless pathway activity. *Development* 134:2807–2813.

Peichel, C. L., K. S. Nereng, K. A. Ohgi, B. L. E. Cole, P. F. Colosimo, C. A. Buerkle, D. Schluter, and D. M. Kingsley. 2001. The genetic architecture of divergence between threespine stickleback species. *Nature* 414:901–905.

Pennings, P. S., and J. Hermisson. 2006. Soft sweeps III: the signature of positive selection from recurrent mutation. *PLoS Genetics* 2:e186.

Phillips, P. C., and S. J. Arnold. 1989. Visualizing multivariate selection. *Evolution* 43:1209–1222.

Piatigorsky, J., W. E. O'Brien, B. L. Norman, K. Kalumuck, G. J. Wistow, T. Borras, J. M. Nickerson, and E. F. Wawrousek. 1988. Gene sharing by δ-crystallin and argininosuccinate lyase. *Proceedings of the National Academy of Sciences of the United States of America* 85:3479–3483.

Piatigorsky, J., and G. Wistow. 1991. The recruitment of crystallins: new functions precede gene duplication. *Science* 252:1078–1079.

Piatigorsky, J., and G. J. Wistow. 1989. Enzyme/crystallins: gene sharing as an evolutionary strategy. *Cell* 57:197–199.

Pigliucci, M. 1996. How organisms respond to environmental changes: from phenotypes to molecules (and vice versa). *Trends in Ecology and Evolution* 11:168–173.

Pigliucci, M. 2001. *Phenotypic Plasticity: Beyond Nature and Nurture.* The Johns Hopkins University Press, Baltimore.

Prud'homme, B., N. Gompel, A. Rokas, V. A. Kassner, T. M. Williams, S. D. Yeh, J. R. True, and S. B. Carroll. 2006. Repeated morphological evolution through *cis*-regulatory changes in a pleiotropic gene. *Nature* 440:1050–1053.

Przeworski, M., G. Coop, and J. D. Wall. 2005. The signature of positive selection on standing genetic variation. *Evolution* 59:2312–2323.

Ptashne, M. 2004. *A Genetic Switch, Phage Lambda Revisited*. Cold Spring Harbor Laboratory Press, Cold Spring Harbor, NY.

Ptashne, M., and A. Gann. 1997. Transcriptional activation by recruitment. *Nature* 386:569–577.

Ptashne, M., and A. Gann. 2002. *Genes & Signals*. Cold Spring Harbor Laboratory Press, Cold Spring Harbor, NY.

Raff, R. 1996. *The Shape of Life: Genes, Development, and the Evolution of Animal Form*. University of Chicago Press, Chicago.

Raff, R. A., and T. C. Kaufman. 1983. *Embryos, Genes, and Evolution*. Indiana University Press, Bloomington.

Rana, B. K., D. Hewett-Emmett, L. Jin, B. H. Chang, N. Sambuughin, M. Lin, S. Watkins, M. Bamshad, L. B. Jorde, M. Ramsay, T. Jenkins, and W. H. Li. 1999. High polymorphism at the human *Melanocortin 1 Receptor* locus. *Genetics* 151:1547–1557.

Ratcliffe, O. J., G. C. Nadzan, T. L. Reuber, and J. L. Riechmann. 2001. Regulation of flowering in *Arabidopsis* by an *FLC* homologue. *Plant Physiology* 126:122–132.

Rees, J. L. 2003. Genetics of hair and skin color. *Annual Review of Genetics* 37:67–90.

Rendel, J. M. 1945. Genetics and cytology of *Drosophila subobscura*. II. Normal and selective matings in *Drosophila subobscura*. *Journal of Genetics* 46:287–302.

Riehle, M. M., A. F. Bennett, and A. D. Long. 2001. Genetic architecture of thermal adaptation in *Escherichia coli*. *Proceedings of the National Academy of Sciences of the United States of America* 98:525–530.

Roch, F., and M. Akam. 2000. *Ultrabithorax* and the control of cell morphology in *Drosophila* halteres. *Development* 127:97–107.

Roff, D. A. 2003. Evolutionary quantitative genetics. Pp. 145–151 *in* B. Hall and W. Olson, eds. *Keywords and Concepts in Evolutionary Developmental Biology*. Harvard University Press, Cambridge, MA.

Rogers, R. L., T. Bedford, and D. L. Hartl. 2009. Formation and longevity of chimeric and duplicate genes in *Drosophila melanogaster*. *Genetics* 181:313–322.

Ronshaugen, M., N. McGinnis, and W. McGinnis. 2002. Hox protein mutation and macroevolution of the insect body plan. *Nature* 415:914–917.

Rosati, B., M. Dong, L. Cheng, S. R. Liou, Q. Yan, J. Y. Park, E. Shiang, M. Sanguinetti, H. S. Wang, and D. McKinnon. 2008. Evolution of ventricular myocyte electrophysiology. *Physiological Genomics* 35:262–272.

Rowland, M. 1988. Management of γHCH/dieldrin resistance in mosquitoes— A strategy for all insects? Pp. 495–500 in *Proceedings of the British Insecticides and Fungicides Conference*. British Crop Protection Council, Brighton.

Rozowski, M., and M. Akam. 2002. *Hox* gene control of segment-specific bristle patterns in *Drosophila*. *Genes & Development* 16:1150–1162.

Rupp, S., E. Summers, H. J. Lo, H. Madhani, and G. Fink. 1999. MAP kinase and cAMP filamentation signaling pathways converge on the unusually large promoter of the yeast *FLO11* gene. *The EMBO Journal* 18:1257–1269.

Sagee, S., A. Sherman, G. Shenhar, K. Robzyk, N. Ben-Doy, G. Simchen, and Y. Kassir. 1998. Multiple and distinct activation and repression sequences mediate the regulated transcription of *IME1*, a transcriptional activator of meiosis-specific genes in *Saccharomyces cerevisiae*. *Molecular and Cellular Biology* 18:1985–1995.

Savage, S. A., M. R. Gerstenblith, A. M. Goldstein, L. Mirabello, M. C. Fargnoli, K. Peris, and M. T. Landi. 2008. Nucleotide diversity and population differentiation of the *Melanocortin 1 Receptor* gene, *MC1R*. *BMC Genetics* 9:31.

Scarcelli, N., J. M. Cheverud, B. A. Schaal, and P. X. Kover. 2007. Antagonistic pleiotropic effects reduce the potential adaptive value of the *FRIGIDA* locus. *Proceedings of the National Academy of Sciences of the United States of America* 104:16986–16991.

Scheiner, S. M. 1993. Genetics and evolution of phenotypic plasticity. *Annual Review of Ecology and Systematics* 24:35–68.

Scortecci, K. C., S. D. Michaels, and R. M. Amasino. 2001. Identification of a MADS-box gene, *FLOWERING LOCUS M*, that represses flowering. *The Plant Journal* 26:229–236.

Scott, D. A., R. Carmi, K. Elbedour, G. M. Duyk, E. M. Stone, and V. C. Sheffield. 1995. Nonsyndromic autosomal recessive deafness is linked to the *DFNB1* locus in a large inbred Bedouin family from Israel. *The American Journal of Human Genetics* 57:965–968.

Sheldon, C. C., J. E. Burn, P. P. Perez, J. Metzger, J. A. Edwards, W. J. Peacock, and E. S. Dennis. 1999. The *FLF* MADS box gene: a repressor of flowering in *Arabidopsis* regulated by vernalization and methylation. *The Plant Cell* 11: 445–458.

Sheldon, C. C., D. T. Rouse, E. J. Finnegan, W. J. Peacock, and E. S. Dennis. 2000. The molecular basis of vernalization: the central role of *FLOWERING LOCUS C (FLC)*. *Proceedings of the National Academy of Sciences of the United States of America* 97:3753–3758.

Sherman, P. W. 1988. The levels of analysis. *Animal Behaviour* 36:616–619.

Shindo, C., M. J. Aranzana, C. Lister, C. Baxter, C. Nicholls, M. Nordborg, and C. Dean. 2005. Role of *FRIGIDA* and *FLOWERING LOCUS C* in determining variation in flowering time of *Arabidopsis*. *Plant Physiology* 138: 1163–1173.

Simmons, M. J., and J. F. Crow. 1977. Mutations affecting fitness in *Drosophila* populations. *Annual Review of Genetics* 11:49–78.

Simpson, G. G., and C. Dean. 2002. *Arabidopsis*, the Rosetta stone of flowering time? *Science* 296:285–289.

Small, S., A. Blair, and M. Levine. 1992. Regulation of *even-skipped* stripe 2 in the *Drosophila* embryo. *The EMBO Journal* 11:4047–4057.

Small, S., R. Kraut, T. Hoey, R. Warrior, and M. Levine. 1991. Transcriptional regulation of a pair-rule stripe in *Drosophila*. *Genes & Development* 5:827–839.

Smith, R., E. Healy, S. Siddiqui, N. Flanagan, P. M. Steijlen, I. Rosdahl, J. P. Jacques, S. Rogers, R. Turner, I. J. Jackson, M. A. Birch-Machin, and J. L. Rees. 1998.

*Melanocortin 1 receptor* variants in an Irish population. *Journal of Investigative Dermatology* 111:119–122.

Stauber, M., H. Jackle, and U. Schmidt-Ott. 1999. The anterior determinant *bicoid* of *Drosophila* is a derived *Hox* class 3 gene. *Proceedings of the National Academy of Sciences of the United States of America* 96:3786–3789.

Stauber, M., A. Prell, and U. Schmidt-Ott. 2002. A single *Hox3* gene with composite *bicoid* and *zerknullt* expression characteristics in non-cyclorrhaphan flies. *Proceedings of the National Academy of Sciences of the United States of America* 99:274–279.

Steiner, C. C., J. N. Weber, and H. E. Hoekstra. 2007. Adaptive variation in beach mice produced by two interacting pigmentation genes. *PLoS Biology* 5:e219.

Stenseth, N. C., W. Falck, O. N. Bjornstad, and C. J. Krebs. 1997. Population regulation in snowshoe hare and Canadian lynx: asymmetric food web configurations between hare and lynx. *Proceedings of the National Academy of Sciences of the United States of America* 94:5147–5152.

Stern, D. L. 2000. Perspective: Evolutionary developmental biology and the problem of variation. *Evolution* 54:1079–1091.

Stern, D. L. 2003. The *Hox* gene *Ultrabithorax* modulates the shape and size of the third leg of *Drosophila* by influencing diverse mechanisms. *Developmental Biology* 256:355–366.

Stern, D. L., and V. Orgogozo. 2008. The loci of evolution: How predictable is genetic evolution? *Evolution* 62:2155–2177.

Stinchcombe, J. R., C. Weinig, M. Ungerer, K. M. Olsen, C. Mays, S. S. Halldorsdottir, M. D. Purugganan, and J. Schmitt. 2004. A latitudinal cline in flowering time in *Arabidopsis thaliana* modulated by the flowering time gene *FRIGIDA*. *Proceedings of the National Academy of Sciences of the United States of America* 101:4712–4717.

Sturtevant, A. H. 1915. Experiments on sex recognition and the problem of sexual selection in *Drosophila*. *The Journal of Animal Behavior* 5:351–366.

Sucena, E., I. Delon, I. Jones, F. Payre, and D. L. Stern. 2003. Regulatory evolution of *shavenbaby/ovo* underlies multiple cases of morphological parallelism. *Nature* 424:935–938.

Sucena, E., and D. L. Stern. 2000. Divergence of larval morphology between *Drosophila sechellia* and its sibling species caused by *cis*-regulatory evolution of *ovo/shavenbaby*. *Proceedings of the National Academy of Sciences of the United States of America* 97:4530–4534.

Suetsugu, N., F. Mittmann, G. Wagner, J. Hughes, and M. Wada. 2005. A chimeric photoreceptor gene, *NEOCHROME*, has arisen twice during plant evolution. *Proceedings of the National Academy of Sciences of the United States of America* 102:13705–13709.

Takahata, N., K. Ishii, and H. Matsuda. 1975. Effect of temporal fluctuation of selection coefficient on gene frequency in a population. *Proceedings of the National Academy of Sciences of the United States of America* 72:4541–4545.

Tanaka, T., K. Ikeo, and T. Gojobori. 2006. Evolution of metabolic networks by gain and loss of enzymatic reaction in eukaryotes. *Gene* 365:88–94.

Tatsuta, H., and T. Takano-Shimizu. 2006. Genetic architecture of variation in sex-comb tooth number in *Drosophila simulans*. *Genetic Research* 87:93–107.

Ting, C. T., S. C. Tsaur, M. L. Wu, and C. I. Wu. 1998. A rapidly evolving homeobox at the site of a hybrid sterility gene. *Science* 282:1501–1504.

Tishkoff, S. A., F. A. Reed, A. Ranciaro, B. F. Voight, C. C. Babbitt, J. S. Silverman, K. Powell, H. M. Mortensen, J. B. Hirbo, M. Osman, M. Ibrahim, S. A. Omar, G. Lema, T. B. Nyambo, J. Ghori, S. Bumpstead, J. K. Pritchard, G. A. Wray, and P. Deloukas. 2007. Convergent adaptation of human lactase persistence in Africa and Europe. *Nature Genetics* 39:31–40.

Tomoyasu, Y., S. R. Wheeler, and R. E. Denell. 2005. *Ultrabithorax* is required for membranous wing identity in the beetle *Tribolium castaneum*. *Nature* 433:643–647.

Toomajian, C., T. T. Hu, M. J. Aranzana, C. Lister, C. Tang, H. Zheng, K. Zhao, P. Calabrese, C. Dean, and M. Nordborg. 2006. A nonparametric test reveals selection for rapid flowering in the *Arabidopsis* genome. *PLoS Biology* 4:e137.

Tour, E., C. T. Hittinger, and W. McGinnis. 2005. Evolutionarily conserved domains required for activation and repression functions of the *Drosophila* Hox protein Ultrabithorax. *Development* 132:5271–5281.

Turelli, M., and N. H. Barton. 2006. Will population bottlenecks and multilocus epistasis increase additive genetic variance? *Evolution* 60:1763–1776.

Turner, J. R. G. 1981. Adaptation and evolution in *Heliconius*: A defense of Neo-Darwinism. *Annual Review of Ecology and Systematics* 12:99–121.

Van Laere, A. S., M. Nguyen, M. Braunschweig, C. Nezer, C. Collette, L. Moreau, A. L. Archibald, C. S. Haley, N. Buys, M. Tally, G. Andersson, M. Georges, and L. Andersson. 2003. A regulatory mutation in *IGF2* causes a major QTL effect on muscle growth in the pig. *Nature* 425:832–836.

Vuilleumier, S., J. M. Yearsley, and N. Perrin. 2008. The fixation of locally beneficial alleles in a metapopulation. *Genetics* 178:467–475.

Walters, J. W., C. Munoz, A. B. Paaby, and S. Dinardo. 2005. Serrate-Notch signaling defines the scope of the initial denticle field by modulating EGFR activation. *Developmental Biology* 2:415–426.

Wang, W., F. G. Brunet, E. Nevo, and M. Long. 2002. Origin of *sphinx*, a young chimeric RNA gene in *Drosophila melanogaster*. *Proceedings of the National Academy of Science of the United States of America* 99:4448–4453.

Wang, X., and H. M. Chamberlin. 2002. Multiple regulatory changes contribute to the evolution of the *Caenorhabditis lin-48 ovo* gene. *Genes & Development* 16:2345–2349.

Waxman, D., and J. J. Welch. 2005. Fisher's microscope and Haldane's ellipse. *The American Naturalist* 166:447–457.

Weatherbee, S. D., G. Halder, J. Kim, A. Hudson, and S. Carroll. 1998. Ultrabithorax regulates genes at several levels of the wing-patterning hierarchy to shape the development of the *Drosophila* haltere. *Genes & Development* 12:1474–1482.

Weatherbee, S. D., H. F. Nijhout, L. W. Grunert, G. Halder, R. Galant, J. Selegue, and S. Carroll. 1999. *Ultrabithorax* function in butterfly wings and the evolution of insect wing patterns. *Current Biology* 9:109–115.

Weber, K. E. 1990. Increased selection response in larger populations. I. Selection for wing-tip height in *Drosophila melanogaster* at three population sizes. *Genetics* 125:579–584.

Weber, K. E., and L. T. Diggins. 1990. Increased selection response in larger populations. II. Selection for ethanol vapor resistance in *Drosophila melanogaster* at two population sizes. *Genetics* 125:585–597.

Weinreich, D. M., R. A. Watson, and L. Chao. 2005. Perspective: Sign epistasis and genetic constraint on evolutionary trajectories. *Evolution* 59:1165–1174.

Welch, J. J., and D. Waxman. 2003. Modularity and the cost of complexity. *Evolution* 57:1723–1734.

Werner, J. D., J. O. Borevitz, N. Warthmann, G. T. Trainer, J. R. Ecker, J. Chory, and D. Weigel. 2005. Quantitative trait locus mapping and DNA array hybridization identify an *FLM* deletion as a cause for natural flowering-time variation. *Proceedings of the National Academy of Sciences of the United States of America* 102:2460–2465.

West Eberhard, M. J. 1989. Phenotypic plasticity and the origins of diversity. *Annual Review of Ecology and Systematics* 20:249–278.

West-Eberhard, M. J. 2003. *Developmental Plasticity and Evolution.* Oxford University Press, Oxford.

Whitlock, M. C. 2003. Fixation probability and time in subdivided populations. *Genetics* 164:767–779.

Whitlock, M. C., and N. H. Barton. 1997. The effective size of a subdivided population. *Genetics* 146:427–441.

Whitlock, M. C., and R. Gomulkiewicz. 2005. Probability of fixation in a heterogeneous environment. *Genetics* 171:1407–1417.

Whitlock, M. C., P. C. Phillips, F. B.-G. Moore, and S. J. Tonsor. 1995. Multiple fitness peaks and epistasis. *Annual Review of Ecology and Systematics.*

Wichman, H. A., M. R. Badgett, L. A. Scott, C. M. Boulianne, and J. J. Bull. 1999. Different trajectories of parallel evolution during viral adaptation. *Science* 285:422–424.

Wiellette, E. L., and W. McGinnis. 1999. *Hox* genes differentially regulate *Serrate* to generate segment-specific structures. *Development* 126:1985–1995.

Wilczek, A. M., J. L. Roe, M. C. Knapp, M. D. Cooper, C. Lopez-Gallego, L. J. Martin, C. D. Muir, S. Sim, A. Walker, J. Anderson, J. F. Egan, B. T. Moyers, R. Petipas, A. Giakountis, E. Charbit, G. Coupland, S. M. Welch, and J. Schmitt. 2009. Effects of genetic perturbation on seasonal life history plasticity. *Science* 323:930–934.

Wiley, R. H. 1991. Lekking in birds and mammals: behavioral and evolutionary issues. *Advances in the Study of Behavior* 20:201–291.

Wilkins, A. S. 2002. *The Evolution of Developmental Pathways*. Sinauer Associates, Sunderland, MA.

Wilkins, A. S. 2007. Between "design" and "bricolage": Genetic networks, levels of selection, and adaptive evolution. *Proceedings of the National Academy of Sciences of the United States of America* 104 Suppl 1:8590-8596.

Winfree, A. 1964. The scientist as poet. *The Cornell Engineer* November:1–2.

Wistow, G. 1993. Lens crystallins: gene recruitment and evolutionary dynamism. *Trends in Biochemical Sciences* 18:301–306.

Wistow, G. J., J. W. Mulders, and W. W. de Jong. 1987. The enzyme lactate dehydrogenase as a structural protein in avian and crocodilian lenses. *Nature* 326:622–624.

Wittkopp, P. J., B. K. Haerum, and A. G. Clark. 2008. Regulatory changes underlying expression differences within and between *Drosophila* species. *Nature Genetics* 40:346–350.

Wittkopp, P. J., J. R. True, and S. B. Carroll. 2002. Reciprocal functions of the *Drosophila* Yellow and Ebony proteins in the development and evolution of pigment patterns. *Development* 129:1849–1858.

Wolf, J. B., E. D. B. III, and M. J. Wade, eds. 2000. *Epistasis and the Evolutionary Process*. Oxford University Press, Oxford.

Wood, B. 1996. Human evolution. *Bioessays* 18:945–954.

Wray, G. A., M. W. Hahn, E. Abouheif, J. P. Balhoff, M. Pizer, M. V. Rockman, and L. A. Romano. 2003. The evolution of transcriptional regulation in eukaryotes. *Molecular Biology and Evolution* 20:1377–1419.

Wright, S. 1929. Fisher's theory of dominance. *The American Naturalist* 63:274–279.

Wright, S. 1931. Evolution in Mendelian populations. *Genetics* 16:97–159.

Wright, S. 1932. The roles of mutation, inbreeding, crossbreeding and selection in evolution. *Proceedings of the Sixth International Congress of Genetics* 1:356–366.

Wright, S. 1934. Physiological and evolutionary theories of dominance. *The American Naturalist* 68:24–53.

Wright, S. 1951. The genetical structure of populations. *Annals of Eugenics* 15:323–354.

Wright, S. 1955. Classification of the factors of evolution. *Cold Spring Harbor Symposium on Quantitative Biology* 20:16–24D.

Wright, S. 1968. *Evolution and the Genetics of Populations,* vol. 1: *Genetics and Biometric Foundations*. University of Chicago Press, Chicago.

Wright, S. 1980. Genic and organismic selection. *Evolution* 34:825–843.

Yang, Y., and A. Cvekl. 2005. Tissue-specific regulation of the mouse $\alpha A$-crystallin gene in lens *via* recruitment of Pax6 and c-Maf to its promoter. *Journal of Molecular Biology* 351:453–469.

Yokoyama, R., and S. Yokoyama. 1990. Convergent evolution of the red- and green-like visual pigment genes in fish, *Astyanax fasciatus*, and human. *Pro-*

*ceedings of the National Academy of Sciences of the United States of America* 87: 9315–9318.

Yokoyama, S. 2002. Molecular evolution of color vision in vertebrates. *Gene* 300:69–78.

Yokoyama, S., and F. B. Radlwimmer. 1998. The "five-sites" rule and the evolution of red and green color vision in mammals. *Molecular Biology and Evolution* 15:560–567.

Yoo, B. H. 1980a. Long-term selection for a quantative character in large replicate populations of *Drosophila melanogaster*. II. Lethals and visible mutants with large effects. *Genetical Research* 35:19–31.

Yoo, B. H. 1980b. Long-term selection for a quantitative character in large replicate populations of *Drosophila melanogaster*. I. Response to selection. *Genetical Research* 35:1–17.

Zhang, J. 2003. Evolution by gene duplication: an update. *Trends in Ecology & Evolution* 18:292–298.

# INDEX

generations of reduced, 108, 205
human, 108
increasing effective, 203
large, 160, 204, 216
long-term effective, 108, 205
role of, 102, 203
total, 109
population structures, 102, 146–7,
155–6, 160, 166, 169, 171,
190, 206
population subdivision, 205
population thinking, 171
populations, xi–xv, 3–10, 12–13, 20,
22–4, 34, 39–42, 44, 46–7,
49, 59–61, 63–6, 72–3, 75,
78–9, 84–6, 91, 98–9,
101–10, 112–13, 116,
151–3, 155–6, 160–4,
172–4, 189–90, 192,
195–200, 205–7, 211–12,
214–16
diploid, 44
distinct, 141
domesticated, 157
experimental-evolution, 163
fluctuating, 205
founding, 205
human, 12, 108, 111, 186, 190,
206
hypothesized ideal, 103, 203
independent, 192
isolated, 110, 172, 198, 211
laboratory, 153
large asexual, 215
large random-mating, 101
local, 110
morphological evolution, 158
mutagenized, 152
panmictic, 174, 205–6
predator, 202–3
real, 101, 103, 203
reduced effective population size
of, 109, 205
replicate, 85, 163, 199, 216

sexual, 198
structured, 174
subdivided, 151, 161
undivided, 151
predictions, 46–7, 78, 150–1, 158,
169–72, 174, 192–3
protein
coding changes, 121, 215
function, 27, 62, 77, 194, 214
products, xv, 10–11, 80
regions, 11, 89
sequences, 14, 17, 19
proteins, xiv, 2, 10–12, 14, 16–17, 19,
26–30, 33, 51–2, 54, 77,
87–9, 111, 115, 120, 142,
144, 167, 189, 193, 207
ancestral, 87
purifying selection, 13–14, 17–19, 39,
54, 72, 111, 173, 186–7

## Q

Quaternary ice ages, 207

## R

random events, xiii, 43, 185
receptor, 79–83, 87, 90, 110–11, 119,
187, 201, 206, 214
receptor protein, 80–1, 110, 206
Glucocorticoid, 86–90
Mineralocorticoid, 87–8
recessive, 22–3, 34, 36, 46
alleles, 23, 35, 43–4, 187, 191
fitness effects, 43, 46
mutations, 26, 39
recombination, 186, 200
regulation, 18, 86, 120–1, 126, 130,
144, 199
transcriptional, 120
regulatory interactions, 123, 127–8,
130, 142
regulatory network, 122–3, 126–7,
129–30, 142, 144, 147